Chemical warehousing

The storage of packaged dangerous substances

HSE Books

© *Crown copyright 2009*

First published 1992 as *Storage of packaged dangerous substances*
Second edition 1998
Third edition 2004
Fourth edition 2009

ISBN 978 0 7176 6237 1

All rights reserved. No part of this publication may be reproduced, stored in a retrieval system, or transmitted in any form or by any means (electronic, mechanical, photocopying, recording or otherwise) without the prior written permission of the copyright owner.

Applications for reproduction should be made in writing to: Licensing Division, Her Majesty's Stationery Office, St Clements House, 2–16 Colegate, Norwich NR3 1BQ or by e-mail to hmsolicensing@cabinet-office.x.gsi.gov.uk

This guidance is issued by the Health and Safety Executive. Following the guidance is not compulsory and you are free to take other action. But if you do follow the guidance you will normally be doing enough to comply with the law. Health and safety inspectors seek to secure compliance with the law and may refer to this guidance as illustrating good practice.

Photographs taken at Ciba UK plc and DHL Exel Supply Chain sites

Contents

Introduction ...1
Who is this guidance for? ...1
Objectives ...1

Legal requirements ...3
Relevant legislation ...3
Regulators ...5

Hazard identification and risk assessment ...6
Hazards ...6
Hazard classification ...6
Risk assessment ...11

Elimination or reduction of risks ...12
Introduction to the hierarchy of control ...12
Control measures ...12
Mitigation measures ...28

Hazardous area classification ...35

Emergency arrangements ...38
Overall approach ...38
General fire precautions ...38
Emergency procedures ...42
Control of off-site risks ...43
Escape facilities ...44
First aid ...44

Information, instruction and training ...45

Audit and review ...46
Audit ...46
Reviewing performance ...46

References and further reading ...47
References ...47
Further reading ...50
Relevant legislation ...50

Glossary ...52

Further information ...53

Introduction

1 This guidance book sets out control measures aimed at eliminating or reducing risks to people, at work or otherwise, from the storage of packaged dangerous goods.

2 This fourth edition of Chemical warehousing reflects changes to legislation and new sections have been added to reflect changes to industry practice and what chemicals warehouses store.

3 The guidance reflects good practice for the design of new storage facilities. It is accepted that it may be inappropriate or impractical to adopt all recommendations in existing premises. However, the law requires you to make any improvements that are reasonably practicable, taking into account the risks at the premises and the cost and feasibility of the additional precautions.

4 This guidance is specific to the storage of packaged dangerous substances. However, there are other legal health and safety requirements that you should meet, including:

- the Management of Health and Safety at Work Regulations 1999 (as amended);
- the Health and Safety at Work etc Act 1974; and
- the Workplace (Health, Safety and Welfare) Regulations 1992 (as amended);

so it should be used in conjunction with other material published by the Health and Safety Executive (HSE) and the environment agencies, including *Managing contractors: A guide for employers* HSG159,[1] *Storage & handling of drums & intermediate bulk containers* PPG26,[2] *Getting Your Site Right: Industrial and Commercial Pollution Prevention*,[3] and *Warehousing and storage: A guide to health and safety* HSG76.[4]

5 If your chemical warehouse site falls under the requirements of the Control of Major Accident Hazards Regulations 1999 (as amended) (COMAH), you may need to provide supplementary measures to control the risks that are listed within this publication to achieve a level of risk that is as low as reasonably practicable (ALARP). For further guidance on the COMAH Regulations, refer to *A guide to the Control of Major Accident Hazards Regulations 1999 (as amended)* L111.[5] Where warehouses are storing large inventories of dangerous substances, eg flammable liquids, it may be necessary to read this book in conjunction with other guidance such as *The storage of flammable liquids in containers* HSG51.[6]

Who is this guidance for?

6 This book is aimed at anyone who has responsibility for the storage of dangerous substances, regardless of the size of storage facility, as well as inspectors responsible for enforcing health and safety legislation within them. This guidance is aimed at eliminating or reducing risks to people from this type of work activity. There may also be risks to the environment from these activities but they are outside the scope of this guidance. Advice on risks to the environment should be sought from the Environment Agency in England and Wales, the Scottish Environment Protection Agency (SEPA) in Scotland, or the Northern Ireland Environment Agency in Northern Ireland. These agencies are referred to collectively as the 'environment agencies' throughout this book.

7 This guidance provides advice for operators of storage sites for packaged dangerous substances. It applies to transit or distribution warehouses, open-air storage compounds and facilities associated with a chemical production site or end-user. It is limited to the requirements for safe storage – further HSE guidance is available on other topics. **No chemical processing activities should be undertaken within a warehouse environment** to ensure there is no increased risk to people working within.

Objectives

8 The objectives of this publication are to:

- help you assess and reduce the risks associated with the storage of packaged dangerous substances;
- increase awareness of the potential hazards associated with the storage of packaged dangerous substances;

- advise on safe management procedures and precautions to reduce injuries and damage caused by incidents involving the storage of packaged dangerous substances. Risks to health are also important but this document does not cover them in detail;
- give guidance on the appropriate standards for the design and construction of storage areas and buildings used for storing packaged dangerous substances at ambient temperature and pressure; and
- advise on the need for appropriate precautions, maintenance, training and good housekeeping where packaged dangerous substances are stored.

Legal requirements

9 There are two types of health and safety law: those that make specific requirements detailing what you should do to comply with them and those that make general requirements, such as the Health and Safety at Work etc Act 1974. HSE publishes guidance to help you to understand what you have to do to comply with health and safety law. You can find more information on the HSE website: www.hse.gov.uk.

Relevant legislation

10 Paragraphs 11–41 list some of the health and safety legislation relevant to the storage of packaged dangerous substances.

Carriage of Dangerous Goods and Use of Transportable Pressure Equipment Regulations 2007

11 These Regulations are commonly referred to as the 'Carriage Regulations'. They act as one consolidated piece of legislation replacing the previous range of regulations and aim to protect everyone directly involved in the carriage of dangerous goods (such as consignors or carriers) or who may become indirectly involved (such as members of the emergency services and public). They are enforced by HSE, the Office of Rail Regulation, the Vehicle and Operator Services Agency and the police.

Chemicals (Hazard Information and Packaging for Supply) Regulations 2002 (as amended) (CHIP)

12 These Regulations, commonly referred to as 'CHIP', contain requirements for the supply of chemicals. They require suppliers of chemicals to:

- classify them according to their hazards;
- give information about the hazards to the people they supply, both in the form of labels and material safety data sheets (MSDSs); and
- package the chemicals safely.

13 Classifying chemicals according to CHIP requires knowledge of physical and chemical properties, and of the health and environmental effects.

14 As a result of the 2008 CHIP Amendment Regulations, chemical suppliers may need to take account of any large quantities of newly reclassified chemicals they have stored on site at any one time to ensure that they comply with the Control of Major Accident Hazards Regulations 1999 (as amended).

European Regulation on classification, labelling and packaging (CLP)

15 The Globally Harmonised System of Classification and Labelling of Chemicals (GHS) is a United Nations system to identify hazardous chemicals and to inform users about these hazards through standard symbols and phrases on the packaging labels and through material safety data sheets. These new regulations align existing EU legislation to the GHS.

16 The CLP regulations entered into force on 20 January 2009. They replace the current rules on classification, labelling and packaging of substances (Directive 67/548/EEC) and preparations (Directive 1999/45/EC) after a transitional period. The deadline for substance classification according to the new rules will be 1 December 2010 and for mixtures 1 June 2015.

Control of Pesticides Regulations 1986 (as amended)

17 These Regulations, made under the Food and Environmental Protection Act 1985, require pesticides to be approved before they may be sold, supplied, used, advertised or stored. In addition, sites storing more than 200 kg, 200 litres or a similar mixed quantity of a pesticide approved for agricultural use should be under the control of a person holding a recognised certificate of competence.

Control of Major Accident Hazards Regulations 1999 (as amended) (COMAH)

18 The COMAH Regulations implement Council Directive 96/82/EC, known as the Seveso II Directive, as amended by Directive 2003/105/EC (Control of Major Accident Hazards (Amendment) Regulations 2005) and replaced the Control of Industrial Major Accident Hazards Regulations 1984 (CIMAH). The aim of these Regulations is to prevent major accidents involving dangerous substances and limit the consequences to people and the environment of any which do occur.

19 The COMAH Regulations require dutyholders to introduce controls on establishments where dangerous substances are present above certain quantities. These controls may be to a much higher standard than is normally considered adequate or 'good practice'. The requirements placed upon the dutyholder vary depending on the inventory of dangerous substances on site (known as 'top tier' and 'lower tier'). The requirements are significantly increased for top-tier sites. These Regulations apply to establishments rather than individual activities.

20 The COMAH Regulations are jointly enforced by HSE and the environment agencies acting as the Competent Authority.

Control of Substances Hazardous to Health Regulations 2002 (as amended) (COSHH)
21 COSHH requires that employers control exposure to hazardous substances to prevent ill health. Employers should protect employees and others who may be exposed to these substances in the form in which they occur in the work activity. Substances are classified as hazardous under CHIP.

Dangerous Substances (Notification and Marking of Sites) Regulations 1990
22 The purpose of these Regulations is to help the Fire and Rescue Service by ensuring the provision of advance and on-site information on sites containing large quantities of dangerous substances.

23 The Regulations apply to sites containing total quantities of 25 tonnes or more of dangerous substances. They require suitable signs to be erected at access points and at any locations specified by an inspector, and notification to the appropriate fire and enforcing authorities of the presence of any dangerous substance. *Notification and marking of sites. The Dangerous Substances (Notification and Marking of Sites) Regulations 1990. Guidance on Regulations* HSR29[7] gives further guidance.

Dangerous Substances and Explosive Atmospheres Regulations 2002 (DSEAR)
24 The DSEAR Regulations enact the ATEX User Directive (1999/92/EC) and implement the Chemical Agents Directive. The Regulations aim to protect the safety of workers and others who may be at risk from dangerous substances that can cause fire, explosion or any other similar energy-releasing events.

25 The Regulations require employers to carry out a suitable risk assessment for work activities involving dangerous substances, then eliminate or reduce the risks and if necessary carry out a hazardous area classification exercise. Further guidance is available in the DSEAR ACOP series.[8–12]

Electricity at Work Regulations 1989
26 These Regulations impose requirements for electrical systems and equipment, including work activities on or near electrical equipment. They also require electrical equipment that is exposed to any flammable or explosive substance, including flammable dusts, liquids or vapours, to be constructed or protected so as to prevent danger.

27 Further guidance is available in *Memorandum of guidance on the Electricity at Work Regulations 1989* HSR25.[13]

Equipment and Protective Systems Intended for Use in Potentially Explosive Atmospheres Regulations 1996 (as amended) (the EPS Regulations)
28 The EPS Regulations enact the ATEX Equipment Directive (94/9/EC) known as ATEX 95. These Regulations are aimed at manufacturers and suppliers. They apply to equipment, protective systems, safety devices, controlling devices, regulating devices and components for use in potentially explosive atmospheres. They require that the equipment is safe, that it meets essential health and safety requirements, has undergone an appropriate conformity assessment and is affixed with CE marking.

Fire (Scotland) Act 2005
29 In Scotland, this Act and its various amendments implemented the same provisions as the Regulatory Reform (Fire Safety) Order did in England and Wales (see paragraph 39).

Health and Safety at Work etc Act 1974
30 This Act is concerned with the health, safety and welfare of people at work and with protecting those who are not at work (members of the public etc) from risks to their health and safety arising from work activities. The general duties in sections 2–4 and 6–8 of this Act apply to all work activities that are the subject of this guidance book.

31 The Act is enforced either by HSE or by local authorities as determined by the Health and Safety (Enforcing Authority) Regulations 1989. Storage operations are enforced by local authorities unless the main business is the storage of dangerous goods, in which case HSE is the enforcing authority.

Health and Safety (Safety Signs and Signals) Regulations 1996
32 These Regulations bring into force the EC Safety Signs Directive (92/58/EEC) on the provision and use of safety signs at work. They cover various means of communicating health and safety information. They require employers to provide specific safety signs whenever there is a risk that has not been avoided or

controlled by other means, eg by engineering controls and safe systems of work. The Regulations apply to all places and activities where people are employed. However, they exclude signs and labels used in connection with the supply of substances, products and equipment or the transport of dangerous goods.

Management of Health and Safety at Work Regulations 1999 (as amended) (the Management Regulations)

33 The Management Regulations require employers and the self-employed to assess the risks to employees and others who may be affected by their work activities, so that they can decide what measures need to be taken to comply with health and safety law.

34 The Regulations go on to require you to implement appropriate arrangements for managing health and safety, health surveillance (where appropriate), emergency planning, and the provision of health and safety information and training. An Approved Code of Practice L21[14] gives guidance on these regulations.

Notification of Installations Handling Hazardous Substances Regulations 1982 (NIHHS)

35 These Regulations require premises with specified quantities of particular substances to be notified to the local authority. Following the Planning (Control of Major-Accident Hazards) Regulations 1999 and the Planning (Hazardous Substances) Act 1990 and Regulations 1992, the presence of NIHHS Schedule 1 substances and quantities, together with some from COMAH Schedule 1, on, over or under land requires consent from hazardous substances authorities.

Planning (Control of Major-Accident Hazards) Regulations 1999

36 These Regulations enact the land use planning controls as required in the Seveso II Directive (96/82/EC). They are enforced by local authorities within the UK.

37 The Regulations state that where dutyholders have an inventory of a dangerous substance that exceeds a minimum quantity, they are required to apply for hazardous substances consent. Once consent has been granted, the land use planning laws are 'triggered' and the inventory of chemicals on the site will be considered for planning applications in the vicinity.

Provision and Use of Work Equipment Regulations 1998

38 These Regulations aim to ensure the provision of safe work equipment and its safe use. They include general duties covering the selection of suitable equipment, maintenance, information, instructions and training. They also address the need for equipment to control selected hazards. They require employers to ensure that people using work equipment are not exposed to hazards arising from its use.

Regulatory Reform (Fire Safety) Order 2005

39 This legislation, which came into force in England and Wales in October 2006, replaces the Fire Precautions Act 1971 and the Fire Precautions (Workplace) Regulations 1997 (as amended) and revokes the Fire Certificate (Special Premises) Regulations 1976. The Order enshrines the principle of a 'responsible person' ensuring that a fire risk assessment is carried out and fire safety duties are complied with.

40 The Order applies to all persons at work plus all persons lawfully on the premises and those not on the premises but in its vicinity who may be affected by a fire on the premises.

Regulators

41 The Health and Safety (Enforcing Authority) Regulations 1998 define which regulator enforces a workplace. Warehouses are enforced by HSE or the local authority dependent upon use, except those that are subject to the COMAH Regulations, where enforcement is by the Competent Authority (see paragraphs 18–20).

Hazard identification and risk assessment

Hazards

42 A number of hazards may be created when storing packaged dangerous substances. These hazards may affect people working within the storage site, the emergency services in the event of an incident, the general public off site and the environment.

43 In a warehouse, fire is generally considered to be the greatest hazard. This is because many people can be exposed to dangers such as radiated heat, missiles, harmful smoke and fumes. There will also be other hazards within your storage area that you should consider. In rare cases, certain stored substances can undergo violent decomposition when engulfed in flame, and an explosion can result.

44 Common causes of incidents are:

- lack of awareness of the properties of the dangerous substances;
- operator error, due to lack of training and other human factors;
- inappropriate storage conditions with respect to the hazards of the substances;
- inadequate design, installation or maintenance of buildings and equipment;
- exposure to heat from a nearby fire or other heat source;
- poor control of ignition sources, including smoking and smoking materials, hot work, electrical equipment etc; and
- horseplay, vandalism and arson.

Hazard classification

45 Packaged dangerous goods have their own well-defined hazards, often detailed on the material safety data sheet (MSDS), and a specified safe method of storage. Section 15 of the MSDS summarises all relevant hazardous information about a product in terms of the CHIP labelling requirements. However, certain types of packaged dangerous substances may give rise to additional hazards within a warehouse. These different types of dangerous substances should therefore be assessed when considering a risk control strategy to ensure there is sufficient segregation etc. Interaction between different dangerous substances may create additional hazards.

46 Dangerous substances should be received into a chemical warehouse by a competent person who understands all the risks that they pose and can decide on where to store them and how to segregate them, having regard to their physical and chemical properties, the quantities concerned and the sizes of the packages.

47 Most dangerous substances arriving on site will be marked with the carriage labelling or marking system laid out in the Carriage Regulations. These Regulations refer to the system set out within the ADR[15] regulatory regime. This is based on a global classification system, is widely understood in industry and is simple to operate. Most chemical warehouses use this system to classify dangerous substances, although it does not meet the requirements of DSEAR. However, many dangerous goods, such as aerosols, are transported in limited quantity packages and will not be marked with the carriage labelling system, but with the limited quantity mark. This will normally consist of the United Nations (UN) number, or numbers, (identifying the type of dangerous goods) within a framed diamond and preceded by the letters 'UN'. This is shown in black on a white background. Where multiple numbers are required these may optionally be replaced by the letters 'LQ'. Examples of limited quantity marks are shown in Figure 1.

Figure 1 Examples of limited quantity marks

48 From 2009 a new Excepted Quantities package is being introduced. This is for very small quantities in special packages. These will be marked as shown in Figure 2.

49 There is also a second hazard classification system in operation, known as CHIP, that is based on information for supply of the dangerous substance.

* = position for Class or Division
** = place for name of consignee or consignor
(if not shown elsewhere on package)

Figure 2 Marking for Excepted Quantities packages

The classification tests are comparable to those for carriage, but are not the same as they serve a different purpose. A substance may have different hazard indicators for carriage and supply. In these cases the competent person should consider both hazard classification categories. In general, if a substance is within its outer packaging and this has a carriage label then the carriage system should be used. If the substance is removed from the outer packaging then the CHIP labelling should be used. The introduction of the classification, labelling and packaging regulations should make any differences between the two systems obsolete. COMAH and COSHH use the CHIP hazard classification system. This system does not necessarily meet the more specific requirements of DSEAR.

50 The CHIP system generally uses black on orange–yellow danger symbols, with associated signal words as shown in Table 1.

Table 1 CHIP classification system

Used with indication of danger statement(s)	Type of hazards covered
EXPLOSIVE (E)	Explosive and pyrotechnic products
VERY TOXIC (T+) or TOXIC (T) or	Acutely toxic chemicals
[None]	Carcinogens, mutagens and toxic by reproduction (not normally dangerous for carriage)
HARMFUL (Xn) or	Chemicals with a harmful effect (may not be dangerous for transport). Also used for inhalation sensitisers
IRRITANT (Xi)	Non-corrosive chemicals which may cause inflammation. Also used for skin sensitisers
CORROSIVE (C)	Chemicals which may destroy living tissue on contact
DANGEROUS FOR THE ENVIRONMENT	Chemicals that may present an immediate or delayed danger to one or more environmental compartments
EXTREMELY FLAMMABLE or	Liquid substances with very low flashpoint and boiling point, and flammable gases
HIGHLY FLAMMABLE	Chemicals which may become hot and catch fire OR solids which may readily catch fire after brief contact with an ignition source OR liquid chemicals with flashpoint below 21 °C OR which when wet give off flammable gas in dangerous quantities
OXIDISING	Chemicals which give rise to highly exothermic reaction in contact with other combustible substances

Chemicals that have a flashpoint between 21 °C and 55 °C are identified merely by the presence of risk phrase R10 'Flammable'.

51 If goods are in transit, between activities on site, or intended for dispatch at a later time, the competent person should collect relevant information from within the company to determine the hazard classification to allow a storage location to be identified.

52 The hazard classification categories are given in paragraphs 53–80 and Table 2. Note that Class 1 (explosives), Class 6.2 (infectious substances) and Class 7 (radioactive substances) are not considered in this book.

Class 2: Gases, compressed, liquefied, or dissolved under pressure

Figure 3 Labelling for Class 2

53 In addition to the labelling required under the various regulations (the Carriage Regulations, CHIP and DSEAR), cylinders are often colour-coded to provide indication of their contents. European standard BS EN 1089-3:2004[16] relates to the colour-coding of transportable gas cylinders.

54 Minor leaks from cylinders of compressed gases may disperse more readily if the cylinders are stored in the open air. Cylinders of liquefied gases should be stored in an upright position so that any leaks from valves etc will be of vapour or gas rather than liquid. Cylinders kept in an upright position should be secured to prevent toppling.

55 Most types of cylinder will explode if exposed to intense heat, causing a risk of impact to people in the vicinity even if the cylinder contents are non-hazardous. Acetylene cylinders in particular are liable to explode without warning during or for some time after exposure to heat, because of the self-decomposition of the product. They may also explode if dropped or struck forcibly.

56 Where flammable, toxic or asphyxiant gas cylinders are stored in buildings, good ventilation is essential to ensure that minor leaks will disperse safely. When considering storage locations and determining ventilation design criteria, your assessment will need to consider whether the gases concerned are heavier or lighter than air.

Class 3: Flammable liquids

Figure 4 Labelling for Class 3

57 Under the Carriage Regulations, liquids are classified as flammable if they have a flashpoint below 60 °C. However, diesel, gas oil and heating oil (light) (UN 1202) are classified as flammable up to a flashpoint of 100 °C.

58 This definition of flammable liquid includes all liquids classified as flammable, highly flammable and extremely flammable for supply according to CHIP.

59 Flammable liquid fires can grow rapidly once the integrity of the container is breached and the fire will spread quickly when escaping liquid flows from the stored material. If the fire comes into contact with other flammable or oxidising materials, it will significantly increase in size and it will be more difficult to bring under control. Sealed containers may explode if exposed to intense heat. Depending on ground conditions at the time, liquids may travel some distance while a leak remains undetected.

60 Precautions to be taken include storing flammable liquids in a cool, dry place away from sources of ignition and heat, and in securely closed containers specifically designed for the purpose. It is preferable for the store to be in the open air, but in all cases adequate ventilation at high and low level will be needed to disperse any vapours from leaking containers. You should also consider means of controlling spillage – see *The storage of flammable liquids in containers* HSG51.[6]

Class 4

61 This class contains materials with a variety of hazards and physical properties. Some are low melting point solids, or solids which are kept under a protective layer of inert liquid or gas. The types of substance included are described briefly in paragraphs 63–68, under the three recognised divisions of the class.

62 You should obtain advice on each particular type of substance from the supplier. This needs to include information about any special conditions required for safe storage, eg temperature limitations, or sensitivity to impact, friction, impurities or water.

Class 4.1: Flammable solids

Figure 5 Labelling for Class 4.1

63 These are readily combustible solids that can be ignited by brief contact with a source of ignition, or are sensitive to friction, and that will continue to burn after removal of the source of ignition. Examples are matches, firelighters, nitrocellulose and sulphur.

64 Self-reactive substances are included in this division. These may decompose with the evolution of heat and fumes at moderate temperatures. Examples include various azo compounds such as AZDN.

65 Also included in this division are desensitised explosives. These contain sufficient water, solvent or plasticiser to suppress their explosive properties. Care should be taken to ensure that water or solvent-wetted explosives are not able to dry out through inappropriate storage conditions. Examples of desensitised explosives include picric acid and urea nitrate.

Class 4.2: Self-reactive and related substances

Figure 6 Labelling for Class 4.2

66 Pyrophoric (spontaneously combustible) substances have packaging that is designed to exclude air. If air enters a damaged package the substance may start to burn at room temperature or when gently heated. Examples include yellow phosphorus and some metal alkyls.

67 Oxidative self-heating substances may react with the air, raising the temperature to the point at which spontaneous combustion takes place. This is normally a slow process, which can be controlled by restricting the pack size, limiting storage duration, monitoring temperatures or excluding air. Examples include some types of carbon dust and oily natural products.

Class 4.3: Substances dangerous when wet

Figure 7 Labelling for Class 4.3

68 These substances react with water and evolve flammable gases. Fire involving or in the vicinity of such materials should obviously not be tackled with water. Examples include calcium carbide, metal hydrides, powders of reactive metals such as magnesium or aluminium, and the alkali metals such as sodium.

Class 5: Oxidising substances and organic peroxides

69 These are substances which, although they may not in themselves be combustible, can help other materials to burn rapidly even if air is excluded. When heated in a fire, many of these substances decompose and give off oxygen, which can increase the rate of burning with possible catastrophic consequences. It should be noted that organic peroxides do burn, often explosively – see paragraphs 70–73.

70 There have been a small number of incidents where relatively large packaged quantities of these materials, such as ammonium nitrate and sodium chlorate, have been involved in violent explosions when engulfed in fire. Guidance on the maximum stack size for ammonium nitrate is provided in *Storing and handling ammonium nitrate*[17] and for sodium chlorate in *Storage and use of sodium chlorate and other similar strong oxidants* CS3.[18]

Class 5.1: Oxidising substances

Figure 8 Labelling for Class 5.1

71 Most substances classified as oxidising are extremely reactive. They may be solids or liquids. They need to be stored away from flammable materials, so preventing any contamination or any possibility of them becoming involved in a fire. They may be stored with other similar strong oxidising agents, provided they are compatible.

Class 5.2: Organic peroxides

Figure 9 Labelling for Class 5.2   Old label (acceptable until end of 2010)

72 Organic peroxides are a particularly reactive type of oxidising substance. They may be solids, liquids or pastes, and have one or more of the following properties:

- liable to explosive decomposition;
- burn rapidly and intensely even in the absence of oxygen;
- sensitive to impact or friction;
- react dangerously with other substances;
- decompose at comparatively low temperatures; and/or
- cause spontaneous ignition if spilt onto combustible material.

73 Some organic peroxides may need to be marked with a subsidiary explosion risk label. Organic peroxides need to be stored separately from flammable, corrosive and toxic materials. Advice is given in *The storage and handling of organic peroxides* CS21.[19] Specific advice and information on particular organic peroxides can be obtained from the material safety data sheets or the supplier.

Class 6: Toxic substances

Figure 10 Labelling for Class 6

74 The main risk from toxic substances during storage is failure of containment. Appropriate pre-planning can minimise the consequences of isolated punctured drums or burst packages.

75 However, in the event of fire, such protection is likely to be compromised by the failure of many containers due to the effects of flame and heat. As well as posing an immediate threat to anybody in the vicinity, eg firefighters, the toxic substance can also be spread large distances in the plume of smoke or it may be washed into watercourses by firefighting operations.

76 The precautions necessary to control these risks depend on the quantities of toxic substances involved, their degree of toxicity and their persistence in the environment.

77 Toxic substances vary widely in the hazards they create. During storage, the main hazard is from acute exposure due to loss of containment, rather than from the chronic effects that arise from low-level long-term exposure. Any labelling under the Carriage Regulations will give basic advice on the primary hazards and precautions, but material safety data sheets will need to be consulted for more detailed information.

Class 8: Corrosive substances

Figure 11 Labelling for Class 8

78 Dangerous substances may be classified as corrosive because they burn the skin or otherwise harm anyone coming into contact with them. Many corrosive substances will also react with incompatible or unsuitable packaging or metals, eg storage racking or process plant. Leaking corrosive substances may damage the packaging of other dangerous substances, creating further leaks. Do not assume that different corrosive substances can be safely stored together. In many cases they can react with one another and may give rise to greater hazards than the individual substances. Read the available safety data to determine which substances can be stored together.

Class 9: Miscellaneous dangerous substances

Figure 12 Labelling for Class 9

79 This class includes such diverse substances as asbestos and those that are dangerous to the environment (eg PCBs), lithium batteries, and air bag inflators. You need to consider the specific characteristics of any material in this class before accepting it for storage.

80 The above classifications can be used to assist in segregating dangerous substances – see Table 2. **Segregation is one of the most important risk-control measures in storage.**

Risk assessment

81 DSEAR and the Management Regulations require employers to assess the risks to workers (and others who may be affected by their work or business) which may arise because of the presence of dangerous substances within the workplace.

82 In completing the assessment, employers should consider the hazards arising from their work activity and the risks to employees (or others) and take steps to control these risks.

83 This book details issues you need to consider when undertaking your risk assessment and helps you identify control measures to implement concerning the storage of dangerous substances.

How to carry out a risk assessment

84 Risk assessment can be broken down into five steps:

Step 1   Identify the hazards
Step 2   Decide who might be harmed and how
Step 3   Evaluate the risks and decide on precautions
Step 4   Record your findings and implement them
Step 5   Review your assessment and update if necessary

85 For more information on risk assessment see *Five steps to risk assessment*[20] and the DSEAR[12] or Management Regulations[14] Approved Codes of Practice and Regulations.

86 The hierarchy of control is discussed in paragraphs 93–101 and the preferred options for controlling the risks presented by dangerous substances listed.

Factors to consider when undertaking a risk assessment

87 When assessing risks from dangerous substances, there are a number of factors you should consider. These are specified in regulation 5(2) of DSEAR and include:

- the hazardous properties of the dangerous substance(s);
- safety information provided by the supplier of the dangerous substance, including information contained in any relevant safety data sheet;
- the circumstances of the work, eg the amount of dangerous substances involved in storage and transport activities;
- particular activities which may present a high level of risk such as maintenance or movement of transport;
- the effect of any measures already in place or taken as a result of DSEAR;
- the likely presence of explosive atmospheres and the need for hazardous area classification;
- the likelihood that ignition sources, including electrostatic discharges, will be present;
- the scale of the anticipated effects of a fire or an explosion;
- any places which are or can be connected via openings to places in which explosive atmospheres may occur; and
- any other information you may need to complete the risk assessment.

88 Employers should regularly review their risk assessments and revise them as significant changes arise. Such changes would include the quantity or nature of substances on site, or changes to management or work equipment.

89 Employers who employ five or more people are required, by the Management Regulations, to record the significant findings of their risk assessment. However, it is good practice to record your findings in any case as soon as practicable and make them available to all relevant employees and regulators.

90 Employers should involve their employees or representatives or both in the risk assessment process. They will often be in a good position to help identify what happens in practice during a work activity.

91 It is recommended that the means adopted to control risks from the storage of dangerous packaged substances be written into the company safety policy. A periodic review of these matters should be carried out, particularly if storage conditions change.

92 Further guidance on these factors is available in the DSEAR Approved Code of Practice L138.[12]

Elimination or reduction of risks

Introduction to the hierarchy of control

93 Section 2 of the Health and Safety at Work etc Act 1974 imposes a general duty on employers to ensure, so far as is reasonably practicable, the health, safety and welfare of all their employees. Section 3 of the Act imposes similar duties on employers towards those not employed by them but who may be affected by their activities.

94 The term 'so far as is reasonably practicable' has been interpreted by the courts as allowing cost, as well as time or trouble, to be taken into account as factors to be set against the risk. 'Reasonably practicable' means that you have to take action to control health and safety risks in your workplace, except where the cost (in terms of time and effort as well as money) of doing so is grossly disproportionate to the reduction in the risk.

95 DSEAR expands on this and describes a hierarchy of control measures in relation to dangerous substances and explosive atmospheres. This sets out the priority given to risk control measures and the order in which employers should consider them.

96 Regulation 6(1) of DSEAR requires employers to ensure that risks to employees (and others who may be at risk) are eliminated or reduced so far as is reasonably practicable.

97 Regulations 6(2) of DSEAR requires that preference be first given to substituting the dangerous substance with a different one, or substituting a new or modified work process to eliminate or reduce the risk. Where risks cannot be completely eliminated by substitution, regulation 6(3) requires employers to use a combination of control and mitigation measures to ensure the safety of employees and others.

98 The order in which risk control measures should be considered – the hierarchy of control – is therefore:

- elimination;
- substitution;
- control; and
- mitigation.

99 In a chemical warehouse it may not always be practicable to eliminate or substitute the materials being stored, although it may be possible with substances used as part of the storage operation, such as cleaning materials. Where risks cannot be completely averted through elimination or substitution, an employer should use a combination of control and mitigation measures to ensure safety. **If the correct storage conditions cannot be met for particular dangerous substances, then they should not be permitted on the site.**

100 Examples of types of control and mitigation measures employers may wish to consider within the chemical warehousing environment are further explained in paragraphs 102–218.

101 Operators of chemical storage sites that fall within the scope of COMAH are required to take all measures necessary to prevent major accidents and limit their consequences to people and the environment. The principles of the 'hierarchy of control' may be used to demonstrate that this has been achieved.

Control measures

102 The measures discussed here are designed to control the risks from the storage of packaged dangerous substances. Employers should consider these options as part of the risk assessment required by the Management Regulations and DSEAR. When assessing the risks they will need to consider whether the existing control measures are adequate and suitable. Similarly they should consider what measures need to be included in the design of any new installation.

Risk management

103 At all sites where packaged dangerous substances are to be stored, the management need to consider the risks created and the means adopted to control such risks. The storage of multi-hazard goods together is a high-risk activity demanding high-level management considerations.

104 It is recommended that individual risk management policies be developed for all warehouses or other

premises used to store packaged dangerous substances. The degree of detail in these policies is clearly dependent on various factors, for example:

- quantities stored;
- specific hazards of the materials; and
- location of warehouse(s).

105 A senior member of staff should be directly responsible for safe warehousing operations. Safety management needs to be a key responsibility of the position. It is important that this person should be responsible for the identification, assessment, handling and storage of all the dangerous goods held on site. Clearly this person (or people) needs to be competent to do the job, and should be adequately trained and have sufficient knowledge.

106 Written operating procedures need to be developed covering matters such as selecting storage locations, dealing with spillages, and security arrangements etc. Liaison with the enforcing authorities and the emergency services may be appropriate. Remember that arrangements will be needed for the control of visitors or contractors.

107 If the quantity of dangerous substances on site exceeds the qualifying inventory for COMAH, then you are required, among other things, to inform the Competent Authority. Operators of COMAH sites are required to take all measures necessary to prevent major accidents based on the principle of reducing risk to ALARP. Further guidance may be found in *A guide to the Control of Major Accident Hazards Regulations 1999 (as amended)* L111.[5]

Receipt of goods

108 Employers should make sure they know what materials are being received into the warehouse before they arrive on site. When materials arrive the consignment paperwork should be checked as well as the actual material being delivered and the integrity of the goods and packaging, eg check for leaks.

109 Any substances arriving on site that cannot be identified, or where other problems exist, should not be sent into the storage area. You should have procedures in place for handling such substances and contacting the supplier for help. You may need to store the material in a remote place while this procedure is undertaken. If a satisfactory response from the supplier is not received the material should be removed from site. All staff should be trained and familiar with these procedures.

110 With the exception of those transported in limited quantities, dangerous goods should arrive with either carriage labelling as described in paragraph 47 or supply labelling as described in paragraph 49. The former will be the simplest guide to how the goods should be stored but the latter may also be used as a guide.

111 Substances that have been transported should comply with the Carriage Regulations. Materials that have come into the country from the EU will be labelled using the ADR scheme,[15] which uses the same classification system as the Carriage Regulations. ADR is a European system, so goods entering the chemical warehouse may be marked in this way.

Figure 13 A goods receipt area

112 However, the specific rules of the Carriage Regulations only require the outer container of packaged goods to be marked up according to this classification system. For packaged goods, the internal packages will normally be marked using the CHIP Regulations, unless being imported from outside the EU in which case they will have to be re-labelled to the CHIP system upon receipt. CHIP requires that:

- the hazard of the substance has been classified;
- the substance is suitably labelled with specific risk and safety phrases, and packaged accordingly; and
- information, in the form of a material safety data sheet, is available.

Although a material safety data sheet is only required to be available to the end user, in practice it could be used by an intermediary to inform storage conditions.

113 Duties under COSHH and COMAH are based on the classification system arising from CHIP, so this should be taken into account when undertaking COSHH assessments or determining whether the COMAH Regulations apply. Sometimes packages arriving onto site will be marked using a combined labelling system for both CHIP and the Carriage Regulations.

Separation and segregation of dangerous goods

114 Before goods arrive on site they should be assessed as part of a receipt procedure to determine the hazards they pose. From this a decision can be made as to where they should be stored within the chemical warehouse in accordance with local segregation policies. Incompatible materials should where practicable be segregated in the reception area, even if they were delivered together, and moved to the storage area as soon as possible. Where necessary interim storage areas should be assigned.

115 Your segregation policy should cover the potential for ignition or escalation of an incident. Often it is not the dangerous substance that is the first material to be ignited in a fire; in many cases it is other materials, such as discarded packaging, pallets or rubbish, ignited by a discarded cigarette or a spark from poorly controlled hot work. Such materials should be removed from the warehouse – or placed in a suitable separate compartment. Pallets outside the warehouse should not be stacked against the wall unless it is fire-resisting. Similarly, dangerous substances inappropriately stored within general storage can significantly increase the severity of a fire. This then increases the dangers to on-site personnel, the emergency services and people off-site, as well as to the environment. Areas should be clearly marked

Figure 14 Chemical drum store

to show the types of substances that can be stored in them.

116 The intensity of a fire, or its rate of growth, may be increased if incompatible materials are stored together. For example, oxidising agents will greatly increase the severity of a flammable liquid fire, or the storage of packaged free-flowing flammable powder, especially stored at height, can increase the fire spread if the packaging fails. Furthermore, a fire may grow and involve dangerous substances which, in themselves, are not combustible. In this way, toxic materials can be widely dispersed in the smoke plume or carried in the firefighting water, leading to potential consequences off site to people or the environment or both. To avoid such escalation dangerous substances should be stored in dedicated warehouses or suitable compartments of warehouses.

117 Your segregation policy should be used to prevent such escalation. If you store a large range of multi-hazard stock, it may not be feasible to assess each substance individually and store it accordingly. The various classification and labelling systems described earlier can be used to greatly simplify the assessment.

118 Where a substance is likely to degrade during storage, you will need to consult the suppliers concerning the possible hazardous effects of such degradation. Ask them to provide you with:

- the remedial actions to be taken;
- the recommended storage conditions;
- maximum storage times; and
- inspection frequencies.

119 Table 2 gives recommendations for the segregation of dangerous substances according to their hazard classification. It uses the classification system described within the Carriage Regulations. This system is globally recognised, relatively simple to operate and well understood by industry. It uses nine classes and where a material has more than one classification there is an agreed hierarchy to determine the most appropriate classification. The table excludes Class 1 (explosives), Class 6.2 (infectious substances) and Class 7 (radioactive substances).

120 The segregation advice set out in Table 2 does not take account of chemical incompatibilities. In some cases, different substances that are shown as compatible in the table may react together. You should also consult the material safety data sheets and other available sources for reactivity data to determine whether it is safe to store them together. This particularly applies to many corrosive substances

Figure 15 Storage of packaged dangerous goods in racking

Chemical warehousing

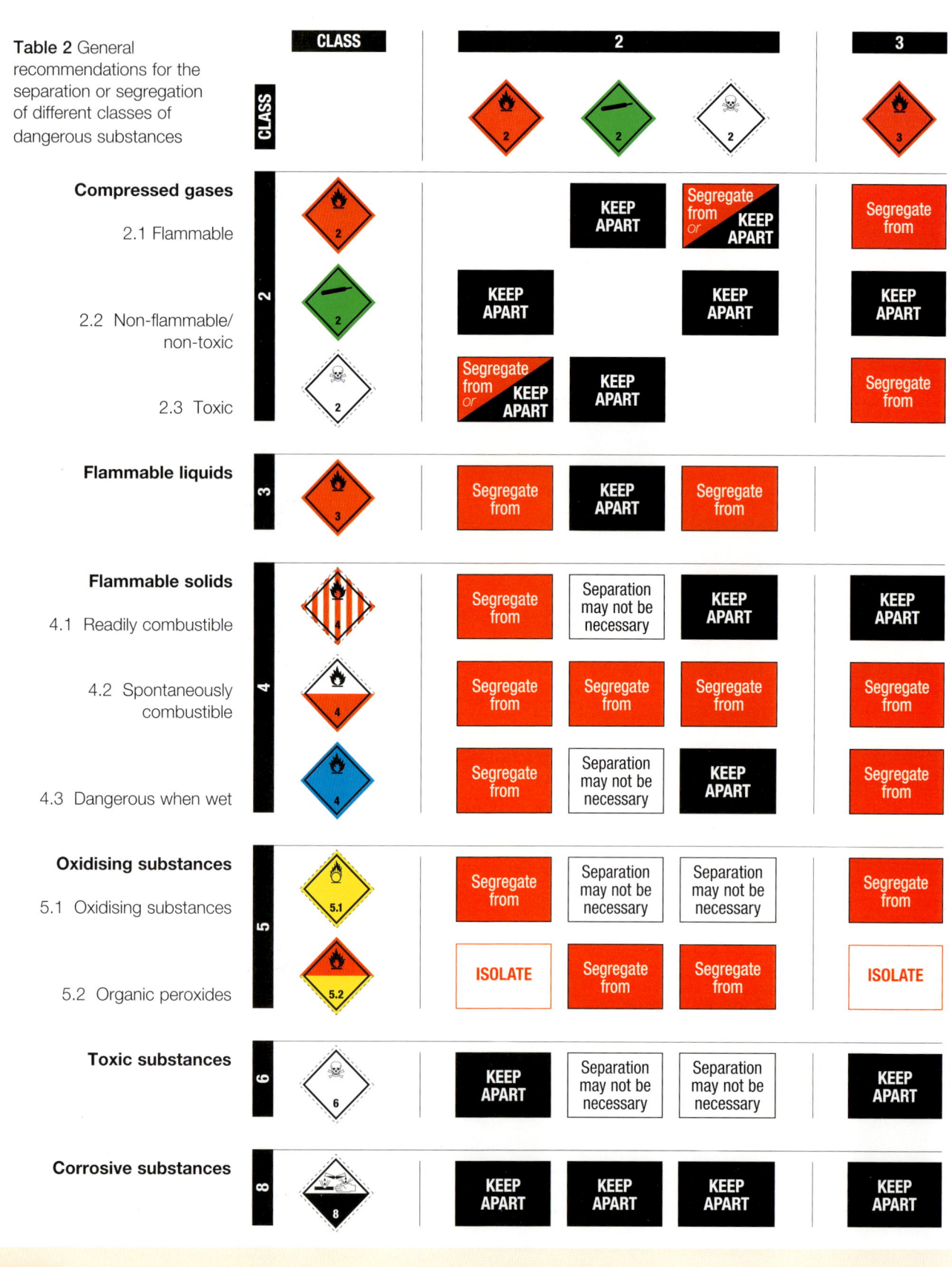

Table 2 General recommendations for the separation or segregation of different classes of dangerous substances

CLASS		2 — 2.1 Flammable	2 — 2.2 Non-flammable/non-toxic	2 — 2.3 Toxic	3
Compressed gases 2.1 Flammable			KEEP APART	Segregate from or KEEP APART	Segregate from
2.2 Non-flammable/non-toxic		KEEP APART		KEEP APART	KEEP APART
2.3 Toxic		Segregate from or KEEP APART	KEEP APART		Segregate from
Flammable liquids 3		Segregate from	KEEP APART	Segregate from	
Flammable solids 4.1 Readily combustible		Segregate from	Separation may not be necessary	KEEP APART	KEEP APART
4.2 Spontaneously combustible		Segregate from	Segregate from	Segregate from	Segregate from
4.3 Dangerous when wet		Segregate from	Separation may not be necessary	KEEP APART	Segregate from
Oxidising substances 5.1 Oxidising substances		Segregate from	Separation may not be necessary	Separation may not be necessary	Segregate from
5.2 Organic peroxides		ISOLATE	Segregate from	Segregate from	ISOLATE
Toxic substances 6		KEEP APART	Separation may not be necessary	Separation may not be necessary	KEEP APART
Corrosive substances 8		KEEP APART	KEEP APART	KEEP APART	KEEP APART

Segregate from: These combinations should not be kept in the same building compartment or outdoor storage compound. Compartment walls should be imperforate, of at least 30 minutes fire resistance and sufficiently durable to withstand normal wear and tear. Brick or concrete construction is recommended. An alternative is to provide separate outdoor storage compounds with an adequate space between them.

Separation may not be necessary: Separation may not be necessary, but consult suppliers about requirements for individual substances. In particular, note that some types of chemicals within the same class, particularly Class 8 corrosives, may react violently, generate a lot of heat if mixed or evolve toxic fumes.

ISOLATE: This is used for organic peroxides, for which dedicated buildings are recommended. Alternatively, some peroxides may be stored outside in fire-resisting secure cabinets. In either case, adequate separation from other buildings and boundaries is required.

Chemical warehousing

4			5		6	8
◇4.1	◇4.2	◇4.3	◇5.1	◇5.2	◇6	◇8
Segregate from	Segregate from	Segregate from	Segregate from	ISOLATE	KEEP APART	KEEP APART
Separation may not be necessary	Segregate from	Separation may not be necessary	Separation may not be necessary	Segregate from	Separation may not be necessary	KEEP APART
KEEP APART	Segregate from	KEEP APART	Separation may not be necessary	Segregate from	Separation may not be necessary	KEEP APART
KEEP APART	Segregate from	Segregate from	Segregate from	ISOLATE	KEEP APART	KEEP APART
	KEEP APART	Segregate from	Segregate from	Segregate from	KEEP APART	Separation may not be necessary
KEEP APART		KEEP APART	Segregate from	ISOLATE	KEEP APART	KEEP APART
Segregate from	KEEP APART		KEEP APART	Segregate from	Separation may not be necessary	Separation may not be necessary
Segregate from	Segregate from	KEEP APART		Segregate from	KEEP APART	KEEP APART
Segregate from	ISOLATE	Segregate from	Segregate from		KEEP APART	KEEP APART
KEEP APART	KEEP APART	Separation may not be necessary	KEEP APART	KEEP APART		Separation may not be necessary
Separation may not be necessary	KEEP APART	Separation may not be necessary	KEEP APART	KEEP APART	Separation may not be necessary	

| KEEP APART | Separate packages by at least 3 m in the storeroom or storage area outdoors. Materials in non-combustible packaging that are not dangerous substances and that present a low fire hazard may be stored in the separation area. This standard of separation should be regarded as a minimum between substances known to react together readily, if that reaction would increase the danger of an escalating incident. | Segregate from or KEEP APART | The lower standard refers to the outside storage of gas cylinders. Where non-liquefied flammable gases are concerned, the 3 m separation distance may be reduced to 1 m. | Where a particular material has the properties of more than one class, the classification giving the more onerous segregation should be used. |

in Class 8, which may react together to produce heat or toxic gases. Examples are:

- acids/hypochlorites – generate chlorine gas;
- acids/cyanides – generate hydrogen cyanide gas;
- acids/alkalis – generate heat; and
- acids/sulphides – generate hydrogen sulphide.

121 Generally, the segregation of acids from other substances will go some way to ensuring incompatible substances are not stored together. The extent of such incompatibility problems is reduced because damage to two packages would need to occur before any reaction can take place. Also, mixing and reaction is likely to be slow if both incompatible components are solids.

122 The miscellaneous dangerous substances in Class 9 and the other dangerous substances in the Carriage Regulations have quite varied properties, and no general advice can be given regarding segregation. You will need to obtain this advice from the supplier.

Storage location

123 If you are considering the location of a new warehouse or outdoor storage compound for storing dangerous substances, then part of your risk assessment will consider, in the case of an incident, the effect of the substances on neighbouring property or populations. Certain sectors of the population are considered more vulnerable than others, eg children in schools, patients in hospitals or residents of retirement homes.

124 If you are already operating at existing premises, your risk assessment will help you decide the quantities or types of materials which can be stored so as not to impose a significant risk on neighbouring populations.

125 The location of new buildings with respect to boundaries is controlled under Building Regulations administered by the local authority. While the specified separation distances reduce the likelihood of fire spreading to neighbouring property, they do not take account of the higher levels of thermal radiation that occur with certain highly flammable materials. The following guidance gives separation distances for some of these materials:

- *The storage of flammable liquids in containers* HSG51;[6]
- *Storage of Full and Empty LPG Cylinders and Cartridges*;[21]
- *Energetic and spontaneous combustible substances: Identification and safe handling* HSG131;[22]
- *Safe use and storage of cellular plastics* HSG92;[23]

Figure 16 Separation of chemicals within a chemical storage area

- *Storage and handling of industrial nitrocellulose* HSG135;[24]
- *Building Regulations 2000 Approved Document B: Fire Safety*;[25]
- *The Building (Scotland) Regulations 2004 Procedural Guidance and Technical Handbook*.[26]

These will typically be greater than those required under the Building Regulations. In every case, the greater separation distance should be adopted.

Means of access/egress

126 There are many different designs and locations of warehouses or outdoor storage compounds. However, in all cases access to the storage area is necessary to carry out various day-to-day operations. The standards applicable for new buildings are covered in guidance made under the Building Regulations, *Building Regulations 2000 Approved Document B: Fire Safety*.

127 You should also remember that access is important in emergencies. The access to the store, and through any site boundary fencing, needs to be adequate for the rapid deployment of firefighting equipment by the local Fire and Rescue Service. The access also needs to be from more than one point, as an incident may make one of the means of access unusable. If the conditions on site are congested, you may need to consider traffic movement schemes, eg speed restrictions and one-way systems.

128 Obviously, access for the Fire and Rescue Service during an incident is paramount and hence this access should be available at all times. If access of unauthorised vehicles is allowed or parking is not controlled, then access by the Fire and Rescue Service may not be possible. These aspects will have been considered before granting planning permission for new buildings.

129 Just as important as access to the warehouse or storage compound are escape routes from the stores for use in an emergency, particularly involving fire. However, means of escape in case of fire form only part of the general fire precautions that are required under the Regulatory Reform (Fire Safety) Order 2005 in England and Wales and the (Fire Scotland) Act 2005 in Scotland. These regulations are enforced by the local fire and rescue authorities, who should be consulted for detailed advice.

Handling and transport

130 Containers should be stacked in a safe manner that facilitates handling operations. The stack design should allow any leaking container to be quickly seen, easily removed and appropriately dealt with. 205-litre metal drums and similar containers are normally stacked no more than four high and preferably on pallets. Drums stored on their sides need to be prevented from moving by suitable chocks. Compressed gas cylinders should either be stored horizontally or secured to prevent toppling; in the case of liquefied petroleum gas (LPG) and acetylene it should be the latter. Containers should not be stacked so as to obstruct ventilation openings or means of escape in case of fire. Stacks should be at least 0.5 m below electric lights. *Storage & handling of drums & intermediate bulk containers* PPG26[2] gives some useful advice.

131 You need to ensure that the most appropriate mechanical handling equipment is used. This is clearly dependent on the types of packages encountered and how they are stored. Specialist fork-lift trucks may be needed to operate in narrow aisle areas, which will

Figure 17 IBCs stored on warehouse racking

require further training. Improvised arrangements for the movement of packages may lead to accidents, damaged packaging and spillage of the contents. Palletised goods need to be secured to prevent accidental movement during handling operations.

132 Vehicles containing packaged dangerous goods need to be parked in a safe place during loading or unloading. Access to and from the site, and to particular storage buildings or compounds, needs to be considered. Your risk assessment will also need to consider the possibility of a collision with a vehicle that may result in the spillage of dangerous goods. Where separation distances to the boundary of the premises apply to permanent storage compounds, you are recommended to maintain these distances wherever possible. For instance, avoid parking loaded vehicles in these areas for long periods.

Operations

133 The store should not be used for activities where spills are more likely, eg dispensing, mixing, and processing. Such operations should be carried out in a separate area, and in a way that reduces spills and dangerous releases. The risk from such operations is greatest with flammable materials, particularly liquids. In these cases, operations should be carried out within a fire-resisting enclosure that is suitably bunded to contain any spills or in a safe place in the open air or in a separate building. This control measure should allow some protection against a fire in the operations area spreading to involve stored goods. *Getting Your Site Right: Industrial and Commercial Pollution Prevention*[3] provides information on precautions that should be taken during delivery, and drainage arrangements for delivery areas.

134 Make sure that any racking installed is properly designed and constructed so as to be stable, and inspected and maintained to ensure that it remains sound. The maximum loading should not be exceeded. Consider how you can load the racking to avoid generating unstable stacks, eg by loading empty racks from the bottom up. You will also need to consider the level at which you store goods in relation to their hazard classification, eg if substances leak

Figure 18 Narrow aisle fork-lift truck in operation

Figure 19 An automated narrow aisle fork-lift truck in operation

from a high-level rack onto a lower-level rack, will this increase the risk of fire?

135 Your risk assessment should also consider vehicle movements in the store. The supports and racking structure may require protection against vehicle impact.

136 Some warehouses are not racked and goods are simply stored in block stacks. Stack sizes may need to be limited to restrict the severity of any fire. In these cases you should set standards for the maximum stack size and height. Stacking heights should be limited so that the lowest layer of packages will not be overloaded and the stability of the pile not endangered. You can get advice on stacking heights from the supplier of the material and, in particular, the design stacking capacities of containers such as drums and intermediate bulk containers (IBCs). Note: IBCs for dangerous goods include the UN IBC marking – an indication of the stacking capability.

Security

137 Physical control measures can minimise the risks of fire or explosion, but these can be defeated if trespassing or tampering is allowed to take place. Your security arrangements, both during the working day and outside normal hours, need to consider the possibility of arson and vandalism. During the working day it should not be possible for an unauthorised person to enter the storage area unchallenged. One way of achieving this is to keep the storage area locked, with access to the keys being restricted to authorised people.

138 The standard of security required will depend, among other things, on the consequences of a major fire. Intruder alarms, security patrols etc may be considered appropriate, but you should not forget other simple precautions such as maintaining fences and external walls. Broken windows and missing construction panels and sheets should be fixed. It is through openings of this type that fires can be deliberately started or unauthorised entry into the store can occur. Stacks of pallets or empty drums up against the building may assist unauthorised access and can also act as the fuel source for an arsonist.

139 Where security fencing is installed around the storage area, its design should take full account of the general fire precautions required.

Control of ignition sources

140 It is important that where an explosive atmosphere may be present all sources of ignition be controlled. This should be considered as part of your DSEAR risk assessment. There are many possible sources of

Figure 20 A gatehouse

ignition that should be considered, examples include:

- smoking and smokers' materials;
- maintenance work, particularly involving hot work;
- electrical supplies and equipment;
- hot surface ignition sources, eg storage close to hot pipes or light fittings;
- arson;
- heating systems incorporating open flames;
- warehouse vehicles, and battery charging facilities;
- LPG-fuelled shrink-wrapping machines;
- radio frequency energy sources, eg mobile phones;
- static electricity; and
- spontaneous combustion, eg if rags or paper contaminated with oil or paint are not properly disposed of.

141 Make sure you maintain control over all potential sources of ignition at sites storing dangerous substances. Some examples of the precautions that you can take are given in paragraphs 142–157.

Smoking and smokers' materials
142 Smoking and smoking materials have caused fires in the past. Smoking is now banned in places of work that are enclosed or substantially enclosed. Smoking and smoking materials should also be prohibited in unenclosed chemical storage areas or wherever there is likely to be an explosive atmosphere or risk of fire.

Maintenance work, particularly involving hot work
143 A permit-to-work system should be used to control any hot work. Precautions to be taken before, during and after the work include:

- clearing, as far as is reasonably practicable, all flammable or combustible materials away from the work area;
- checking for combustible material on one side of a partition or wall when work is to take place on the other side;
- having suitable fire extinguishers at hand and maintaining a careful watch for fire during the work;
- protecting combustible material that cannot be cleared by providing suitable screens or partitions;
- examining the area thoroughly for some time after the work has finished to make sure there is no smouldering material present; and
- as a sensible precaution, stopping all hot work by a safe period before the end of the working day.

144 Burning or welding work at high level is particularly hazardous as hot fragments may travel a considerable distance and still be capable of igniting flammable or heat-sensitive materials.

Electrical supplies and equipment
145 The Electricity at Work Regulations 1989 require any electrical equipment, fixed or portable, to be correctly designed, installed and maintained. For fixed installations, guidance on the correct design, installation and periodic inspection and testing to control the risk of fire and electric shock is given in BS 7671:2008 *Requirements for electrical installations*.[27] Links to other guidance, including guidance on the maintenance and use of portable appliances, is available at www.hse.gov.uk/electricity. For new storage facilities it is good practice to install main switch and distribution boards in a separate fire-resisting room located at the main store entrance, or preferably accessible directly from the outside.

146 If electrical equipment is installed within the store, eg lighting, then you need to ensure that the equipment is suitable for its intended use, correctly positioned (eg ensuring that easily ignitable materials are not stored close to it) and adequate preventative maintenance carried out. Similarly, power cables should be kept clear of any area where they might be attacked by a leak of corrosive substance or mechanically damaged.

Figure 21 Designated smoking area away from the chemical store

147 It is good practice to turn off all non-essential electrical equipment, preferably at the main isolation switch, outside normal working hours and when the store is unoccupied for long periods of time.

148 The DSEAR risk assessment will identify the hazardous areas and the classification of the zones. Using this, you can then establish the standard required for the electrical equipment sited in the hazardous areas. The area classification standards require special precautions for the construction, installation and use of equipment to control ignition sources. DSEAR requires that the category of equipment and protective systems should be consistent with the zoning and requirements set out in the EPS Regulations, unless the DSEAR risk assessment finds otherwise. This also applies to portable equipment such as hand-held radios and mobile phones. Advice on selecting, installing and maintaining explosion-protected electrical equipment is given in BS EN 60079-14:2003.[28]

149 It is recommended that you control the use of unauthorised electrical equipment (such as radios, heaters or kettles) in the store. There have been instances when this type of equipment has caused a fire. Such equipment is normally brought into the store from employees' homes once its use at home has ceased. It is likely therefore to be old and in a poor state of repair, and will not have been maintained.

Heating systems
150 Occasionally storage buildings or internal stores containing dangerous goods are heated. In this case the heating system should not be an ignition source. The use solely of indirect heating can achieve this. Examples include radiators fed remotely by hot water pipes, or indirectly fired gas or oil appliances (ie those which take the air for combustion from a safe place and exhaust the products of combustion to the outside air). Electrically heated radiators that comply with BS EN 60079-14:2003 may be used. In all cases the heating system should be protected against the build-up of flammable residues on hot surfaces. Certain solid substances, such as AZDN, have defined safe storage temperatures above which they will decompose, often with catastrophic results. Where maximum safe storage temperatures are identified, ensure that no heated surfaces above that temperature are present around the stored substance.

Shrink-wrapping operations
151 Ideally, heat shrink-wrapping operations should not be carried out in the store. They should take place either in a separate building or in a specifically designed bay within the building. The type of bay necessary will depend on the risk associated with the materials in the store and those being wrapped.

Figure 22 Hot air heating system

Stretch-wrapping is preferred to heat shrink-wrapping as the risk of fire is greatly reduced, although there are a few occasions where it cannot replace the latter. Suitable and sufficient first-aid and firefighting equipment should be readily available wherever heat shrink-wrapping is carried out. After the package has been heat shrink-wrapped it should be removed to a well-ventilated, unenclosed area to ensure that any smouldering packaging is not encouraged to develop into a fire.

Protection of vehicles
152 Vehicles that have to operate within hazardous areas in storage buildings or areas need to be protected to an appropriate standard to avoid ignition of any explosive atmospheres. *Lift trucks in potentially flammable atmospheres* HSG113[29] provides further advice on the use and protection of lift trucks.

153 Vehicles with petrol or LPG engines should not be parked in the storage area outside normal working hours. Recharging batteries generates hydrogen, a flammable gas. Electric-powered vehicles, such as fork-lift trucks, should be recharged in a bay, separate from the store, with good mechanical ventilation.

Figure 23 Battery charging station separate from the storage area

The type of bay necessary will depend on the risk associated with the materials being stored. This should be assessed as part of your DSEAR risk assessment.

Radio frequency (RF)
154 Transmitted radio frequency (RF) power can act as a potential ignition source, particularly RF power from radio transmission masts and also from mobile phone or CB systems. Large, unearthed conducting structures can act as receiving aerials if they are in the path of RF transmissions, eg crane jibs or metal racking. A spark can be created if an earthed person or object touches the unearthed structure. There is a particular risk of fire or explosion if there is the likelihood of an explosive atmosphere within the vicinity of the metal structure, especially if the atmosphere contains a dangerous substance with a low ignition energy, eg hydrogen. You should, therefore, give some thought to RF ignition hazards when undertaking the DSEAR risk assessment.

155 If you are planning to build a new chemical warehouse, you should consider what RF transmissions are in the vicinity, eg the siting of transmitters or masts, and ensure the appropriate control measures are in place to control any risk. You can obtain this information by contacting the Office of Communications (OFCOM) www.ofcom.gov.uk. Similarly, if a new radio transmission mast is to be erected near your facility, you should assess the risk that the structure may impose on your business and take appropriate action to control or eliminate that risk. For further guidance contact the Health Protection Agency (Radiation Protection Division) www.hpa.org.uk/radiation.

Static electricity
156 The discharge of static electricity may produce sparks of enough energy to ignite some explosive atmospheres and has caused a number of fatalities to date. Flammable liquids have in the past been stored and handled in metal containers. In recent years it has become increasingly common to use plastic containers for a number of sound commercial reasons, including cost, corrosion resistance and reduced weight. However, the use of insulating plastic has led to an increased incidence of static build up through handling and liquid movement.

157 You should consider the risk of ignition from static electricity as part of your DSEAR risk assessment. Several sources of guidance are available.[30–33]

Maintenance and modifications
158 Many incidents occur during or as a result of maintenance activities and repairs. Health and safety law requires that work equipment be maintained in a safe condition. Only personnel who are suitably

qualified and authorised, and who fully understand the hazards, should carry out inspections and maintenance. If you use outside contractors to undertake this work, you should ensure they are competent to carry out the work required.

159 The Health and Safety at Work Act etc 1974, the Management Regulations and DSEAR place duties, to ensure safe working practices, on both the company using the services and the contractor. Guidance is also available on selecting and managing contractors in *Managing contractors: A guide for employers* HSG159.[1]

160 It is essential that no maintenance work be done until:

- the potential hazards of the work have been clearly identified and assessed;
- the precautions needed have been specified in detail;
- the necessary safety equipment has been provided; and
- adequate and clear instruction has been given to all those concerned.

161 In most cases, a permit-to-work system should be used to control those maintenance operations that create a source of ignition or could cause damage to the packages. Permits to work are formal management documents. Only those with clearly assigned authority should issue them. The permit to work states what requirements should be complied with before the permit is issued and before the work

1	**Permit title**
2	**Permit number** Reference to other relevant permits or isolation certificates.
3	**Job location**
4	**Plant identification**
5	**Description of work to be done and its limitations**
6	**Hazard identification** Including residual hazards and hazards introduced by the work.
7	**Precautions necessary** Person(s) who carries out precautions, eg isolations, should sign that precautions have been taken.
8	**Protective equipment**
9	**Authorisation** Signature confirming that isolations have been made and precautions taken, except where these can only be taken during the work. Date and time duration of permit.
10	**Acceptance** Signature confirming understanding of work to be done, hazards involved and precautions required. Also confirming permit information has been explained to all workers involved.
11	**Extension/shift handover procedures** Signatures confirming checks made that plant remains safe to be worked upon, and new acceptor/workers made fully aware of hazards/precautions. New time expiry given.
12	**Hand back** Signed by acceptor certifying work completed. Signed by issuer certifying work completed and plant ready for testing and recommissioning.
13	**Cancellation** Certifying work tested and plant satisfactorily recommissioned.

Figure 25 Example of a permit-to-work form

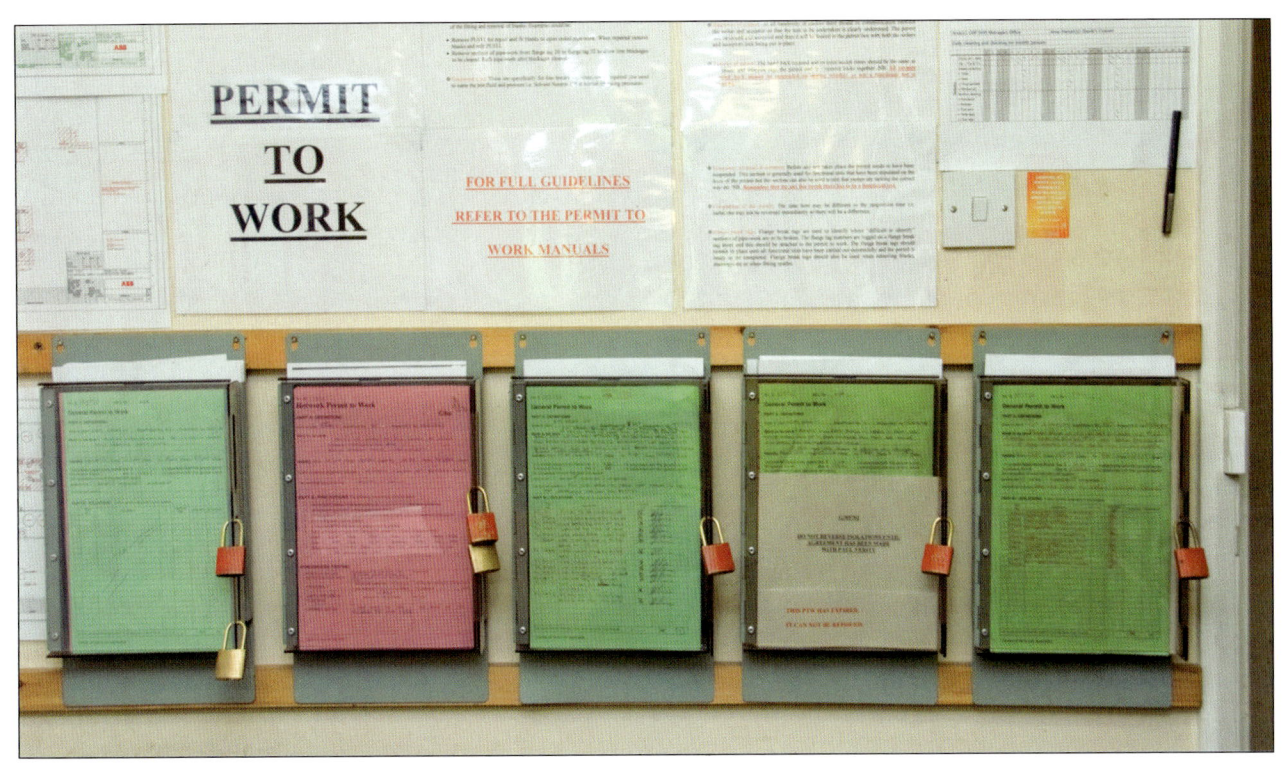

Figure 24 Details of current permits to work in use on site

covered by it is undertaken. Individual permits to work should relate to clearly defined individual pieces of work. Permits to work should normally include (see also Figure 25):

- the location and nature of the work intended;
- identification of the hazards, including the residual hazards and those introduced by the work itself;
- the precautions necessary, eg isolations;
- the personal protective equipment required;
- the proposed time and duration of the work;
- the limits of time for which the permit is valid; and
- the person in direct control of the work.

162 Further advice on permits to work is available in *Guidance on permit-to-work systems. A guide for the petroleum, chemical and allied industries* HSG250.[34]

163 There are some simple controls you can adopt to reduce the risks of fire or explosion during maintenance work. You need to make sure that materials that can burn or be affected by fire are removed from the work area. If it is not reasonably practicable to remove such materials, you should position suitable screens or partitions to protect the hazardous inventory. Once the work has finished, you need to thoroughly inspect the area for about an hour to ensure that there is no smouldering material present. Consideration should also be given to work apparently away from a source of hazard but which may affect other areas, eg during electrical testing work of a fixed electrical system.

Aerosols

164 Some chemical warehouses store aerosols as part of their inventory. Most aerosols use a liquefied flammable gas as a propellant, usually LPG or dimethyl ether, and the risks of storing these should be considered as part of your DSEAR risk assessment.

165 Packages of aerosols, which are usually Limited Quantity packages showing 'UN 1950' in the diamond frame mark, should be inspected on entry into the warehouse to ensure that the contents are not damaged. This could be achieved by visual inspection or by using portable gas detectors while the packages are confined in a transit vehicle en route to the warehouse. Signs to look for include cloudy wrapping, wet packs or strong smells of perfume.

166 A number of destructive fires in aerosol warehouses have occurred as a result of fork-lift truck handling incidents. Dropped pallets, collisions and loose cans have all caused fires when gas released from the damaged cans was ignited by a fork-lift truck. The type of fork-lift truck should be considered carefully as fire statistics show this may be the most likely cause of fire in a chemical warehouse storing aerosols. If the number of aerosol pallet movements is large or other hazardous goods are stored, fork-lift trucks with AC motors and enclosed contactors may be appropriate.

167 Aerosols may rupture if overheated. They should not be stored near heating pipes, hot air vents or in direct sunlight. Repacking operations involving shrink-wrapping pallets or groups of aerosols should not be carried out in the warehouse.

168 Aerosols should not be stored in warehouses that may be subjected to intense heating in the event of an external fire, eg next to a store for highly flammable liquids.

169 Once ignited aerosols can fuel severe fires. Rupturing aerosols generate missiles which make firefighting difficult and can spread fire rapidly. These risks can be reduced using steel mesh barriers around racks of pallets containing aerosols, and restricting the gross quantity stored in each fire-resisting compartment.

170 If possible aerosols should be stored in separate buildings or segregated from other goods by a firewall. In a chemical warehouse this barrier has two

Figure 25 An automated aerosol store

functions: it reduces the risk that fires caused by aerosol handling accidents will spread to involve other hazardous goods; and it reduces the risk that fires caused by handling other hazardous goods will spread to the stored aerosols.

171 Further guidance is available from the British Aerosol Manufacturers' Association (BAMA) in *The BAMA Standard for Consumer Safety and Good Manufacturing Practice: Module 6 Warehousing*.[35]

Intermediate bulk containers (IBCs)

172 The original design use of IBCs was for transport purposes only. However, over recent years the storage of liquids, particularly flammable liquids, in IBCs has significantly increased and many chemical warehouses receive IBCs daily.

173 There are specific risks associated with the storage of IBCs in warehouses, in particular:

- when involved in a fire, IBCs are prone to early failure at the valves or elsewhere. As a consequence, the contents are likely to leak out and fuel the fire, causing rapid escalation. This can happen within minutes of the initial fire taking hold and may lead to the total loss of the warehouse;
- IBCs are liable to degrade when used for long-term storage rather than for transporting materials, which was their original purpose. This may lead to leaking and any subsequent exposure of the contents to an ignition source may result in a fire. There are many mechanisms which may lead to failure of an IBC, eg weathering, stacking on poor surfaces, stacking with no reference to load-bearing ability or certification, use with substances where incompatibilities may arise, use for mixed substances, use for storage of wastes, use as a reaction vessel, damage by vehicles etc. A site inspection procedure should be implemented and any IBCs in a visibly poor condition should be taken out of use. If actually used for transport, a specific inspection regime is set down in ADR;[15]
- IBCs are invariably made of the most cost-effective suitable plastic material, and generally that plastic is non-conductive to electricity. When non-conducting flammable materials are stored in these containers and moved (as in splashing around on the back of a lorry) the surface of the container will become electrostatically charged. This electrostatic charge will either decay with time, or will take the easiest path to earth when it arrives. If that earth path is an employee, they may report having an electric shock when touching the IBC. If there is a small leak of flammable liquid or vapour, the charge may be sufficient to ignite the vapour, escalating into a large uncontrollable fire.

174 You should address these issues when undertaking a specific risk assessment (to comply with DSEAR and the Management Regulations) for storage in IBCs. Where there is a risk of electrostatic charge developing from the materials being carried, appropriate conducting IBCs should be selected. IBCs should also be properly labelled for the Carriage Regulations and CHIP purposes.

175 IBC's containing flammable substances should be stored in bunded areas specifically to reduce the risk of running pool fires. The bund should take into account the volume that could be released from an individual IBC failure and consideration should be given to containing the whole IBC inventory. Where possible, these storage areas should be outside and protected from vehicle damage. It is good practice to stack IBCs only two high within the storage area, unless the stacking height has been confirmed as greater with the manufacturer. Compatibility for stacking should also be checked.

176 Any fire fuelled by substances released from IBCs has the scope for serious health and safety and environmental consequences. COMAH sites in particular should consider this when assessing if they have reduced risks to ALARP.

Figure 26 IBCs stored on warehouse racking

177 There are several different designs of IBCs on the market. Guidance on the hazards associated with storage of the different types of IBCs has been published jointly by the Chemical Business Association (CBA) and the Solvent Industry Association (SIA) in *Guidance for the storage of liquids in intermediate bulk containers*.[36] Further guidance on the storage of IBCs is available in HSE's DSEAR ACOPs,[8-12] *The BAMA Standard for Consumer Safety and Good Manufacturing Practice: Module 6 Warehousing*[34] and the Department for Environment, Food and Rural Affairs' *Groundwater Protection Code*.[37]

Storage of hazardous waste

178 The guidance in this book should be considered as a minimum standard for the storage of new containers of dangerous substances. The storage of hazardous waste is required to comply with the Waste Framework Directive, the Hazardous Waste Directive and in many cases the Integrated Pollution Prevention and Control (IPPC) Directive. For guidance on the environmental requirements you should consult the relevant environment agency (the Environment Agency in England and Wales, the Scottish Environment Protection Agency in Scotland, and the Northern Ireland Environment Agency in Northern Ireland). These agencies and HSE are currently working together to produce guidance that sets out standards that reconcile both environmental and health and safety requirements. This guidance is currently entitled *Proposed Environment Agency, SEPA, NI EHS and HSE joint guidance on the Storage of Hazardous Wastes*.[38] Information about the environmental legislation relating to storage of hazardous waste is also available at www.netregs.gov.uk.

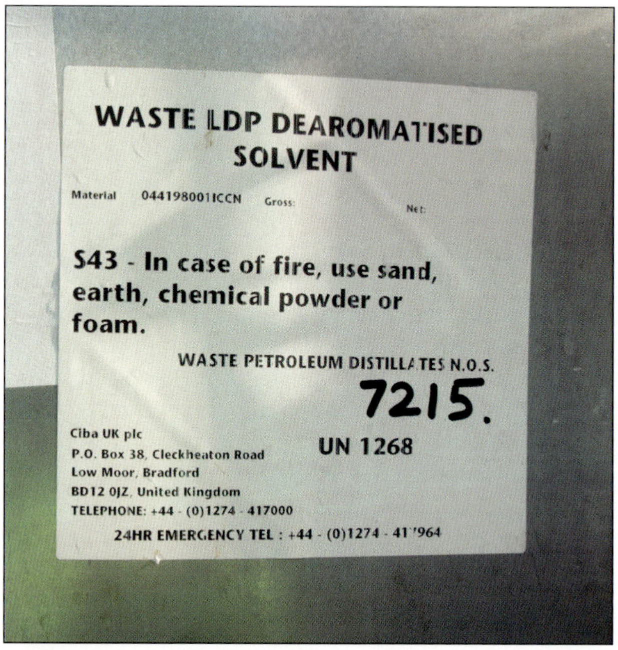

Figure 27 Labelling on a waste container

179 It is important that you are satisfied that you know the contents of waste containers before they are received onto site. You should also ensure that appropriate and unambiguous labels are in place on the containers. The integrity and condition of the storage containers will need to be taken into account in ascertaining appropriate separation distances and segregation within the storage areas.

180 There are specific hazards associated with the storage of hazardous wastes. In particular, waste companies often receive waste in older containers of poorer integrity than new ones. Furthermore, these containers may have previously been used to contain a different dangerous substance and labels may be out of date, inappropriate or ambiguous as to the contents. Where the waste has not been received direct from the initial waste producer and its provenance is uncertain this can also result in inappropriate storage.

Mitigation measures

181 Mitigation measures are at the bottom of the hierarchy of control. They should aim to reduce the harmful physical effects resulting from an incident and to reduce the risk to people and the environment. They are not designed to prevent an incident, rather to limit the consequences arising from one. The list in paragraphs 183–218 is not exhaustive and measures should be selected to be appropriate for the dangerous substances stored within the chemical warehouse. Mitigation measures should be considered as part of the risk assessment for the chemical storage area.

182 Both control and mitigation measures often depend on employees and contractors carrying out the appropriate operating procedures correctly and complying with written or verbal instructions. Therefore, employers should provide employees and contractors with sufficient supervision and training and ensure that operating procedures are correctly followed.

Building construction

183 Storage buildings and outdoor storage compounds for dangerous substances are subject to controls under building and planning legislation. In England and Wales *Approved Document B: Fire safety*[25] sets out standards for fire resistance and compartment size for industrial or storage buildings. The use classes take no account of the specific hazards of the materials being stored, and in some cases, where large quantities of dangerous substances are involved, different or higher standards may be appropriate.

184 In Scotland the building standards are different. A use category specific to the storage of certain types

of dangerous substance is given and more rigorous requirements are imposed.

185 In both cases, the Regulations specify standards for fire resistance, compartment size, means of escape and assistance to the Fire and Rescue Service.

186 Buildings for storing dangerous substances should preferably be constructed of non-combustible materials. Fire-resisting external walls provide some protection against an external fire that may be started deliberately. The use of combustible materials should be considered thoroughly as part of the risk assessment. Buildings may be vulnerable to lightning strike, which may initiate a fire. The building should, where necessary, be provided with suitable protection against lightning. Further guidance on the assessment of lightning risk and measures to reduce it are given in BS EN 62305:2006 *Protection against lightning*.[39]

187 Fire-resisting structures in buildings may be required to prevent the escalation of an incident by preventing the spread of a fire or to protect the means of escape in case of fire. The latter forms part of the general fire precautions required under the Regulatory Reform (Fire Safety) Order 2005 in England and Wales, and the Fire (Scotland) Act 2005 in Scotland. This legislation is enforced, with certain exceptions, by the local fire and rescue authorities, which can advise in matters of general fire precautions. Building regulations also specify standards of fire resistance for certain structures. Scotland has different regulations to those in England and Wales. The regulations are enforced by local authorities. The insurers of your premises may also have requirements for structural fire protection and separation, fire suppression systems and fire compartmentation.

188 In designing storage buildings you need to consider the layout of storage within the warehouse. Some warehouses have an inbuilt store within the main warehouse. This interior store may be used to store hazardous materials, eg highly flammable liquids and gases, aerosols or peroxides, and so the store is required to be fire resisting. Some access into this interior store will be required (see Figure 29).

189 It is better if access to the interior store is available only from outside the main store. This has the advantage of ensuring that the fire resistance between the main store and the interior store is not jeopardised by any access doors being permanently left open. Alternatively the access doors to the interior store may be linked to the fire alarm. Activation of the fire alarm system should result in closure of the fire access doors. It is better to close internal fire doors manually

Figure 28 A purpose-built, bunded chemical drum store

at the end of the working day, rather than depend on the automatic closure devices.

190 To prevent dangerous concentrations of flammable vapours building up within the store as the result of a leak of a highly flammable liquid or gas, the store needs to be adequately ventilated (see Figure 30a). The simplest method of ensuring adequate ventilation is to provide fixed, permanent openings (such as air bricks or louvres) in external walls at high and low levels. If openings are provided on two walls only, a cross flow induced by wind forces is encouraged. Similarly, openings at high and low levels will encourage air circulation by thermal currents. Vehicle access doors may provide sufficient ventilation during the working day, but you also need to consider, in your DSEAR risk assessment, the ventilation provided during out of hours periods.

191 Large buildings may require mechanical ventilation to achieve an adequate air movement. Where this is necessary, it needs to operate constantly. Failure of the ventilation system can be detected by the use of an airflow monitoring device installed in the ventilation ductwork (such as a flow switch or differential pressure switch) and linked to an alarm. In cases of doubt,

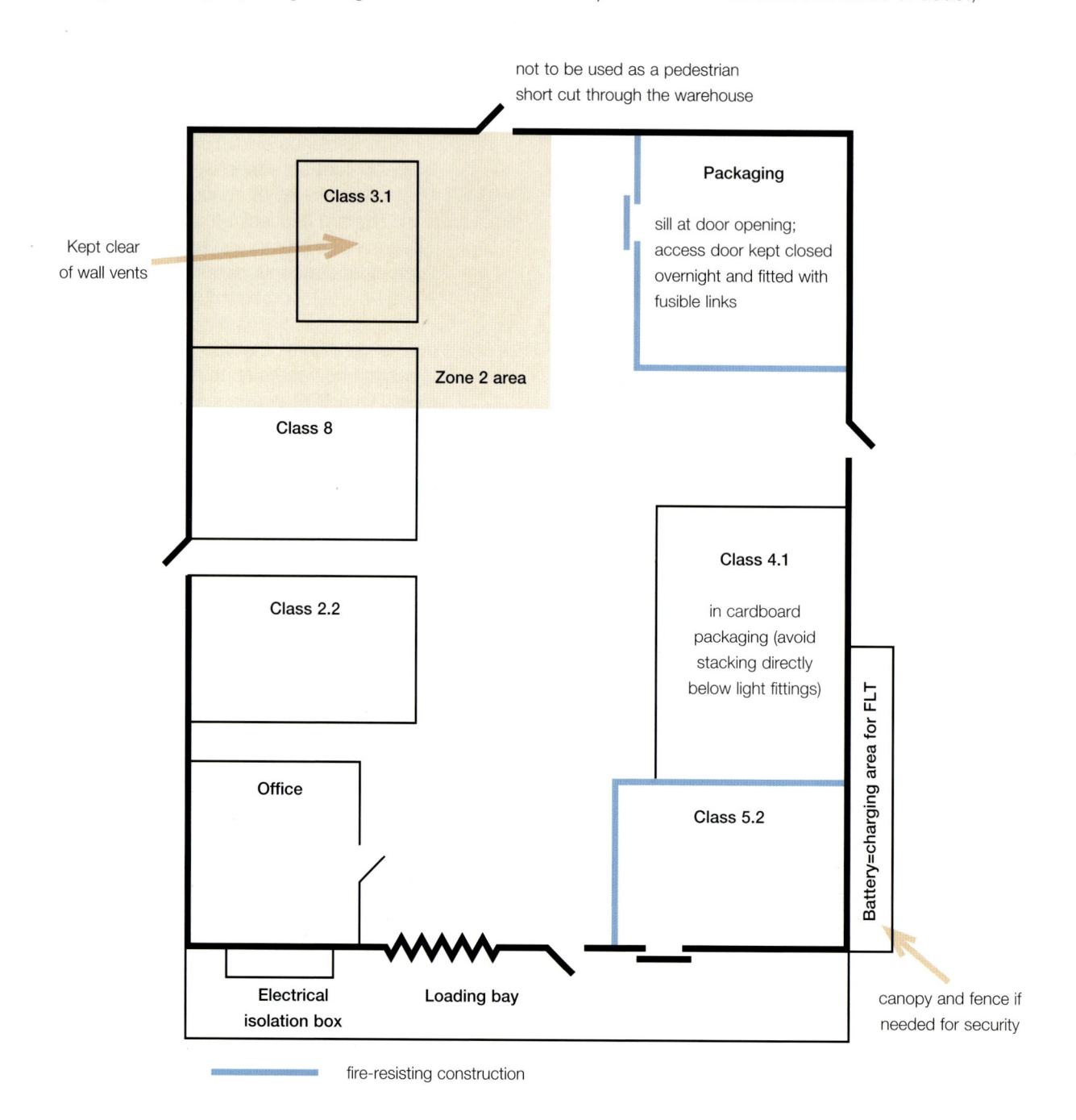

Figure 29 Example of good warehouse layout

measurements may be made of the air change rate actually achieved in a completed building. A competent ventilation engineer should be able to do this.

192 When planning the design of your warehouse, you will need to consider neighbouring properties. This is particularly important for COMAH sites where there may be an off-site risk from your activities. You may need to consider additional safety features, eg firewalls or segregation by distance or both, to reduce the risk to neighbouring properties from your site activities.

Design and construction of packaging and containers

193 Unless stored in tanks or bulk, the main protection against the dangers arising from the storage of dangerous substances is the integrity of the packaging and containers. Individual packages or containers may leak, break or puncture, causing a small escape of material, so arrangements need to be in place to deal with these situations.

194 Both CHIP and the Carriage Regulations require manufacturers, suppliers and distributors to ensure that chemicals are packaged safely.

195 All containers should be designed and constructed to standards suitable for the purpose. They should be robust and have well-fitting lids or tops to resist spillage if knocked over. There are specific standards available for containers and packaging, other than Limited Quantity and Excepted Quantity packages, to comply with transport legislation. Containers need to be of an appropriate UN performance-tested type. A UN-approved container can be identified by referencing the UN mark as specified in ADR 2007.[15] Such containers are also suitable for storage conditions.

196 Where necessary, containers should be protected against corrosion (eg by painting) and against degradation by light, particularly for plastic containers (by suitable shading). In addition, the material from which the containers are made needs to be compatible with the chemical and physical properties of the contents to ensure that no interaction occurs that might cause leakage.

197 If containers are reused, such as for storing commodity chemicals or process waste, they should be individually inspected for damage before refilling and marked as such with arrangements in place for them to be inspected as suitable before reuse. Problems commonly arise from damaged linings to drums, or from corrosion occurring near to the base seams of drums.

Spillage control

198 You should have procedures in place for spillage control, particularly if you are a COMAH site. These should be communicated and practised by your employees, or anyone who is expected to deal with a spillage.

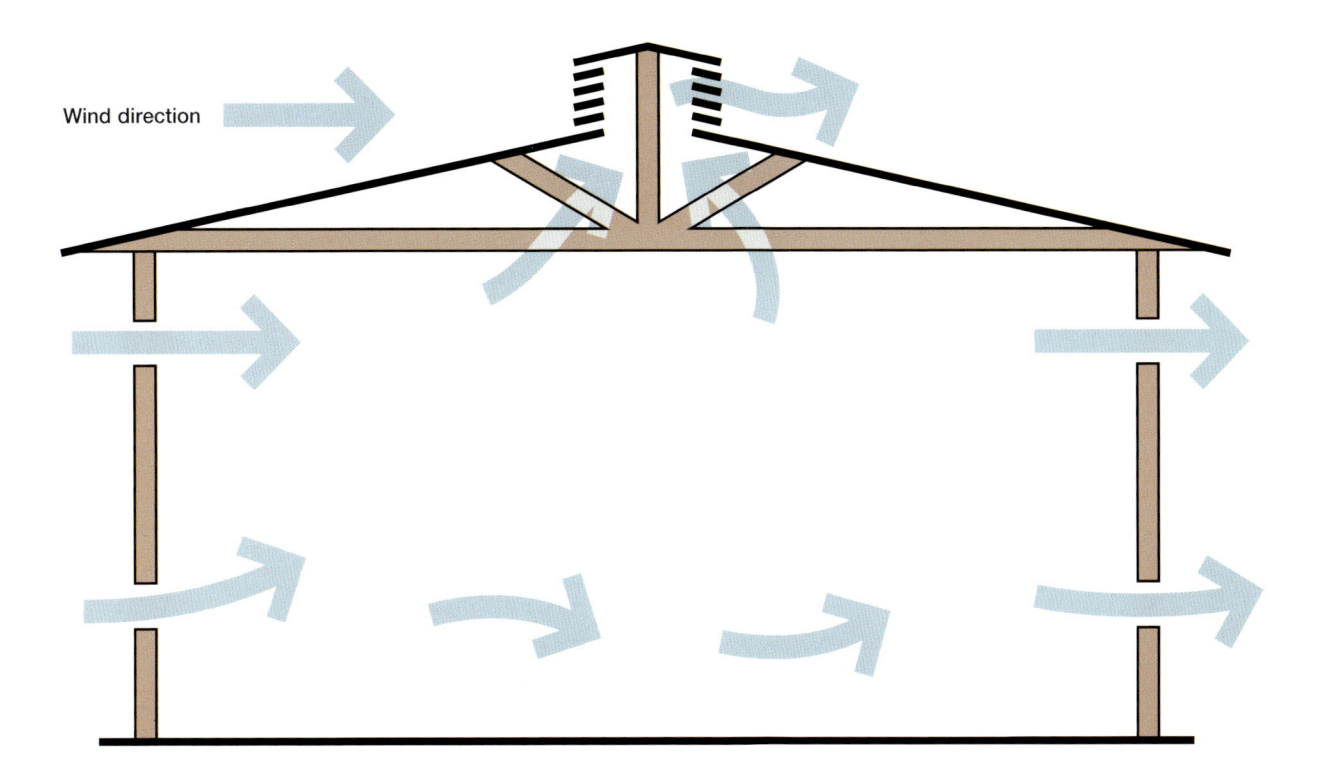

Figure 30 Roof and wall vents together give good natural ventilation

199 It is important to have means of controlling spillages and releases within the storage area to prevent the uncontrolled spread of liquids. A number of control measures are possible. You can provide dry sand, absorbent granules, sealing putties and booms for containing and clearing up small spills where safe to do so. Contaminated materials should then be disposed of safely and appropriately by the use of a registered waste contractor. You should provide a number of spare, clean, empty bags or drums for this purpose. Proprietary salvage drums, sometimes known as overpack drums, are available to hold leaking drums etc. You will need to label the used overbags or salvage drums accordingly. The container used for holding spilled materials should be labelled accordingly. Spillage control materials need to be suitable for use with the spilled materials and readily to hand.

Control of spillages in outdoor storage areas
200 To contain spillages in outdoor storage areas, an impervious sill or low bund can be installed. This should enclose a volume that is at least 110% of the capacity of the single largest container in the bund except in the case of oil storage where 25% of the total volume should be used. Ramps can be provided over the sill to allow fork-lift trucks, pallet trucks etc to access the storage area.

201 The surface of the storage area needs to be impervious and slightly sloped so that any liquid spilt from the containers can flow away to a safe contained place. An alternative method to using a bund is to direct spillages of liquid to another area. This could be to an evaporation area (either within the storage area or separated from it), or via drainage to a remote sump, interceptor or separator. Corrosion of the base of a container can potentially result in leakage of the contents. Good drainage of surface rainwater away from the containers, or the storage of containers on pallets, can reduce the likelihood of this corrosion. It will also reduce the likely contamination of this water and the subsequent disposal problems that would result, as rainwater should be removed to maintain bund capacity. It maybe more cost effective to install a roof or cover over the area.

202 Depending on the hazards of the materials stored, it may be necessary to incorporate a valve-controlled conduit or interceptor pit, or both, in the site drains so that spillages can be retained on site. Fireproof jointing material should be used for any joints in containment systems.

203 Further guidance on bunding and spill control is available from the environment agencies at www.netregs.gov.uk.

Figure 31 Segregated and bunded chemical storage area

204 Combustible materials (including weed growth) need to be excluded from the area surrounding the sill or bund, as their presence increases the fire risk; a one-metre exclusion is considered adequate. If weed growth is controlled by the use of weed killers, you should not use oxidising agents, eg those that contain sodium chlorate.

Control of spillages in buildings
205 Storage rooms or buildings should have floors constructed of materials that are resistant to and compatible with the materials stored. For instance, many acids attack concrete floors, solvents attack bitumen floors, and timber floors impregnated by flammable liquids or oxidising agents such as peroxides are an increased fire risk. Containment of any leaks or releases from containers can be achieved by sloping the floor away from the door, although this may not be possible in warehouses designed for racking, where a sloping floor may compromise racking stability. Leaks can also be contained by providing a sill across the door opening. Typically, such sills are about 150 mm high, and again ramps might be required to allow access for wheeled trolleys, fork-lift trucks etc. The walls up to the height of the sill should also be resistant to and compatible with the material stored. Additional containment may be required if the building's drains link to the site drainage systems.

206 You need to remember that the arrangement of spillage containment and drainage in buildings should take into account the need for material segregation. Liquid spillages should be prevented from running into areas where incompatible materials are stored. This may be achieved in warehouses by installing internal bunded areas, in-rack bunds or drip trays under each pallet and connected to an appropriate sealed drainage system. The bunded volume should be 110% of the single largest receptacle in the bund.

207 Containment and spillage control also needs to take account of the presence of any fire suppression systems. Some lighter-than-water materials can spread by floating on water with such systems. The spillage control should be adequate to cope with the use of the installed fire suppression systems. Internal fire doors are unlikely to prevent the spread of fire from an expanding liquid pool unless sills or appropriate drainage arrangements have been provided at the door opening.

208 These requirements should not be confused with the firefighting water run-off containment (see paragraphs 272–276 on 'Controls for off-site risks') that may be required to prevent the release of materials from the storage area to the environment in the event of an incident. Such run-off containment may need to take account of water that may be applied from both installed systems and manual firefighting. Foam and fire suppression systems, when discharged, may be an environmental hazard and containment needs to be allowed for.

Health precautions
209 Many precautions for reducing fire and explosion risks will also control the health risks. However, some additional measures may be necessary since the concentrations of vapours or dusts capable of damaging human health are usually significantly below explosive levels. COSHH requires employers to prevent or control exposure to harmful substances – guidance is contained in *Control of Substances Hazardous to Health (Fifth Edition): The Control of Substances Hazardous to Health Regulations 2002 (as amended): Approved Code of Practice and guidance*.[40]

210 You need to adopt a safe system of work when dealing with spillages. A number of control measures are possible, and these are described in paragraphs 198–208. Material safety data sheets will detail any specific action to be taken for dealing with spillages. You need to have these available for all the substances stored on site.

211 Spillages need to be cleaned up promptly and the material disposed of safely, in accordance with your site procedures. You should provide precautions against skin and eye contact, such as gloves, protective clothing and goggles. Suitable respiratory protection may be needed during clean-up operations. Substances new to the site should not be handled until suitable personal protective equipment is available.

212 When corrosive materials have been spilt, ensure that employees wear clothing with the necessary resistance to the substance when cleaning up the spillage. This clothing should be removed immediately if contaminated with the dangerous substance. Contaminated clothing should not be sent for cleaning with general laundry or cleaned at an employee's home. It may be cleaned by specialist laundries or disposed of as hazardous waste.

213 Spillages of dangerous substances in a fine, dusty state should not be cleared up by dry brushing. Vacuum cleaners should be used in preference, and for toxic materials, one conforming to type H of British Standard BS 5415[41] should be used. For combustible dusts the vacuum cleaner should not be capable of acting as a source of ignition (see paragraphs 140–149 on 'Control of ignition sources: Electrical supplies and equipment').

Personal protective equipment (PPE)
214 PPE should not be used as a substitute for other methods of risk control. It should always be regarded

as a last-resort means of preventing or controlling exposure to hazards to safety and health. This means that other methods of controlling exposure should be considered before taking the decision to use PPE. In some situations, however, it will be necessary to provide protective equipment.

215 PPE includes both:

- protective clothing, such as overalls, waterproof equipment, gloves, safety footwear, helmets etc; and
- protective equipment, such as eye protectors and ear protectors.

216 Selection of PPE should take into account the demands of the job and the nature of the hazardous substances within the chemical warehouse. Among other things, this will involve considering the physical effort required to do the job, the methods of work, how long the PPE needs to be worn and requirements for visibility and communication. The aim should always be to choose equipment that will give minimum discomfort to the wearer. Uncomfortable equipment is unlikely to be worn properly.

217 There will be considerable differences in the physical dimensions of different workers and therefore more than one type or size of PPE may be needed. There is a better chance of PPE being used effectively if each wearer accepts it. Those having to use PPE should therefore be consulted and involved in the selection and specification of the equipment. There should be no charge made to the worker for the provision of PPE that is used only at work.

218 You should ensure that the PPE you use on site is 'CE' marked and complies with the Personal Protective Equipment Regulations 2002. Further guidance is available in *Personal Protective Equipment at Work (Second edition). Personal Protective Equipment at Work Regulations 1992 (as amended): Guidance on Regulations.*[42]

Figure 32 Signs detailing what PPE to wear within chemical storage area

Hazardous area classification

219 Regulation 7 of DSEAR requires employers to undertake a risk assessment for work activities involving dangerous substances, then eliminate or reduce the risks. Gases, vapours, mists and dusts are known to give rise to explosive atmospheres. If these are present within your chemical warehouse in sufficient quantities to give rise to such a potentially explosive atmosphere that requires special precautions for the protection of people under foreseeable process circumstances (such as a leaking or burst container), you must also complete a hazardous area classification exercise. This identifies where, because of the likelihood of a potentially explosive atmosphere existing, controls over the sources of ignition are required.

220 Hazardous areas or places are classified in terms of zones on the basis of frequency and duration of the occurrence of an explosive atmosphere. Warehouses storing dangerous substances should have a written hazardous zone diagram, which you should retain as part of the documentation to support the risk assessment under regulation 5 of DSEAR. Hazardous zones are defined as follows:

Zone 0   A place in which an explosive atmosphere consisting of a mixture of air with dangerous substances in the form of gas, vapour or mist is present continuously or for long periods or frequently.

Zone 1   A place in which an explosive atmosphere consisting of a mixture of air with dangerous substances in the form of gas, vapour or mist is likely to occur in normal operation occasionally.

Zone 2   A place in which an explosive atmosphere consisting of a mixture of air with dangerous substances in the form of gas, vapour or mist is not likely to occur in normal operation but, if it does occur, will persist for a short period only.

Zone 20   A place in which an explosive atmosphere in the form of a cloud of combustible dust in air is present continuously or for long periods or frequently.

Figure 33 Example of a zoned area

Zone 21   A place in which an explosive atmosphere in the form of a cloud of combustible dust in air is likely to occur in normal operation occasionally.

Zone 22   A place in which an explosive atmosphere in the form of a cloud of combustible dust in air is not likely to occur in normal operation but, if it does occur, will persist for a short period only.

221 To assist in understanding what the definitions mean in a practical way industry bodies have adopted a convention of numerical 'frequency of occurrence' bands that provide a convenient reference point for deciding what constitutes 'long periods' and 'likely to occur'. Typical values are shown in Table 3. These are not mandatory and still require careful judgement to be applied to a particular set of circumstances to ensure the resulting classification provides an adequate degree of safety. Specifically, it may not always be appropriate to equate a secondary-type release with a hazardous zone 2, eg if the release were in a poorly ventilated space this may affect the persistence and duration of the explosive atmosphere.

222 The zone defines the requirements for the selection and installation of certified (fixed and mobile) equipment and protective systems to prevent sources of ignition. The equipment must be selected on the basis of the requirements of the EPS Regulations.

Table 3 Relationship between likelihood of release, type of release and hazardous zone

Probability of flammable concentration of fuel	Type of release	Fuel Liquid, gas or vapour	Dust
Constant or >1000 hours or occurrences per year	Continuous	Zone 0	Zone 20
Intermittent or between 10 and 1000 hours or occurrences per year	Primary	Zone 1	Zone 21
Occasional, or between 1 and 10 hours or occurrences per year	Secondary	Zone 2	Zone 22
Improbable or less than once per year or for less than 1 hour		Unzoned	Unzoned

The hazardous area classification is a purely frequency-based assessment. It estimates the number of occurrences of explosive atmospheres and the duration of those occurrences. The consequences of any ignition and the control and mitigation measures required to operate safely should be dealt with by the risk assessment and control measures based on it made under regulations 5 and 6 of DSEAR.

There is a standardised equipment marking scheme using the 'Ex' equipment logo, shown in Figure 34.

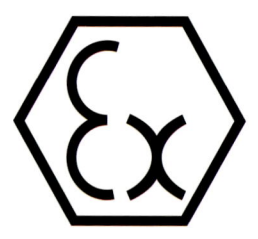

Figure 34 'Ex' equipment logo

223 Where you have classified a hazardous area as a zone, you must warn people they are entering a zone by marking the area or by some other suitable means. The distinctive yellow triangle with an 'EX' logo may be used for this purpose – see Figure 35.

Figure 35 'EX' logo

224 International standard BS EN 60079-10[43] explains the basic principles of area classification for gases and vapours and BS EN 50281:2002[44] for dusts. The DSEAR ACOPs[8–12] also contain guidance on this issue.

225 Judging the extent of any potentially hazardous zone is not easy and can be done in a variety of ways. For most small and medium-sized operations, where materials are in small packages and relatively small quantities, using the generic zone diagrams available in *The storage of flammable liquids in containers* HSG51[6] and the Energy Institute's *Area Classification Code for Installations Handling Flammable Fluids*[45] will provide satisfactory results, as shown in Figure 29.

226 Where the frequency of release may be higher, eg in larger warehouses with larger containers, and/or where there are more product movements, a specific assessment that takes into account the frequency, size and duration of any potential release will need to be made. This may require specialist advice.

227 If you are classifying an existing area as hazardous for the first time you are required to verify that existing equipment is safe, protective systems associated with this area are suitable and that work activities have been designed so they can be carried out safely. The person(s) carrying out the verification must be knowledgeable and experienced in the measures to ensure explosion safety in your work environment. In addition, employees working within these areas must be provided with – and wear – anti-static footwear if your risk assessment indicates that electrostatic discharges could ignite the atmosphere.

228 Sources of ignition have traditionally been thought of as being electrical, but you should also consider all other sources, such as mechanical friction, sparks, air heaters etc (see paragraphs 140–157). Upgrading equipment within a newly zoned area in an existing workplace (to meet the standards defined by the

EPS Regulations) should only be done after more straightforward measures to eliminate the risk have been considered. For example, when the hazardous area has been mapped out it is good practice to attempt to relocate or remove fixed sources of ignition, such as electrical equipment, outside the hazardous area where possible.

229 Portable sources of ignition, such as heaters, kettles etc, should be controlled so that they cannot be used within the defined hazardous areas. There have been instances when this type of equipment, often brought in from home, has caused fires.

230 Vehicles that have to operate within the hazardous areas in storage buildings, or hazardous areas generally, need to be protected to an appropriate standard to avoid the ignition of any flammable vapours released. Specific advice is provided in paragraphs 164–171 for warehouses containing aerosols and *Lift trucks in potentially flammable atmospheres* HSG113[29] provides further advice on the use and protection of lift trucks. There are specific industry-based codes of safe practice (*The BAMA Standard for Consumer Safety and Good Manufacturing Practice: Module 6 Warehousing*[35] and J*oint CBA and SIA guidance for the storage of flammable liquids in sealed packages in specified external storage areas*[46]) that provide additional detail on when protected lift trucks may be required.

231 The CBA/SIA guidance[46] addresses the risks of lift trucks operating within well-ventilated areas outside drum stores and has concluded that special protective measures may not be required for this application provided all other safety measures set out within the code of practice are fully complied with.

Emergency arrangements

Overall approach

232 Much can be done to prevent fire (or other incidents that may have harmful effects), and following the advice in this book should greatly reduce the chances of it occurring.

233 DSEAR and COMAH require employers to assess the likelihood and scale or magnitude of the effects that may result from any foreseeable accident, incident, emergency or other event involving dangerous substances present at the workplace. On the basis of this assessment, employers should put in place appropriate emergency arrangements to safeguard people on their site, mitigate the effects of any such event and restore the situation to normal. Employers will need to determine the degree of intervention appropriate to the circumstances of the emergency.

234 It is not expected that all employers will be able to achieve all mitigation measures to counteract the effects of an incident, although COMAH sites will be expected to adopt more measures than less hazardous sites to reduce the chances or effects of a foreseeable incident.

235 The primary requirement, in an emergency, is that everyone can be evacuated to a place of safety. This book aims to provide guidance on measures that assist with this aim, not limiting consequential loss.

Figure 36 Assembly point

236 Information on emergency arrangements should be made available to employees and their representatives and tested at periodic intervals. Employers may need to provide appropriate training and instruction for employees on these arrangements. Employers will also need to consider which external emergency services may be required, in the event of an emergency, and make them aware of your emergency arrangements. You will need to review these arrangements periodically and revise them if circumstances change at the workplace, eg if you significantly increase the inventory of dangerous substances stored on site. Further guidance is available in the DSEAR ACOP.[12]

237 For chemical warehouses submitting a COMAH safety report, further guidance is available in *Safety Report Assessment Guide: Chemical warehouses*.[47]

General fire precautions

238 If a fire occurs people need to be able to quickly escape and reach a place of safety. The term 'general fire precautions' is used to describe the structural features and equipment provided to achieve this aim. It covers:

- escape routes to fire exits;
- firefighting equipment;
- fixed installations such as water or foam sprinklers or other appropriate media;
- a system of giving warning in the event of fire;
- an efficient arrangement for calling the fire and rescue service; and
- management procedures to ensure that all of the above are available and maintained, and that there is adequate training in their use.

239 You should consider what precautions to adopt when completing your risk assessment under the Regulatory Reform Fire Safety Order 2005 (for England and Wales) or the Fire (Scotland) Act 2005 (for Scotland). Guidance on the application of the former to warehouses can be found in *Guide to Fire Safety in Factories and Warehouses*[48] published by Communities and Local Government, and guidance

relating to the latter can be found in *Practical fire safety guidance for factories and storage premises*,[49] published by the Scottish Executive. Further guidance can also be obtained from your local fire and rescue authority.

Fire detection

240 Outside working hours, or in warehouses that are empty of people for long periods, any outbreak of fire could develop unseen. This could pose a risk to people, both on and off-site, perhaps from smoke containing significant quantities of toxic materials. It may require a means of providing early fire detection. This may be achieved by installing automatic fire detectors that will trigger an alarm, alerting those on site to a fire. They will also, as necessary, warn those in the surrounding area and summon the Fire and Rescue Service. Advice on the selection and installation of suitable equipment is given in BS 5839,[50] where it is recommended that a fire protection engineer who is experienced in the installation of such systems should carry out the work. Again, advice may be obtained from your local fire and rescue authority.

Warning and communication systems

241 Warning and communication systems (including visual and audible alarms) should be provided to alert people to an actual or potential incident involving dangerous substances (see BS 7974[51] and PD 7974[52]). The system should be appropriate to the level of risk presented by foreseeable incidents.

242 There are several types of warning system that can be used. Employers should consider who needs to be alerted and why, the size of the workforce, the quantities and risks of the dangerous substances within the warehouse plus the emergency actions to be carried out when assessing what type of warning system to install.

Firefighting equipment

243 An adequate number of fire extinguishers should be present within the storage area. Their primary purpose is to tackle incipient fires, which often do not involve the dangerous goods, thereby reducing the risk to people and enabling them to make their escape. Anybody expected to use a fire extinguisher should be properly trained. With some types of dangerous substances any attempt to fight a fire may be unwise (eg aerosols), but the ability to a tackle a waste bin or small packaging fire might prevent a serious incident occurring. Further detailed guidance can be obtained from the above-mentioned publications or from your local fire and rescue authority.

244 The extinguishers need to be positioned in conspicuous locations along the escape routes, such that nobody in the storage area needs to travel more than 30 metres to reach one. Unless the location of an extinguisher is self-evident, its position needs to be identified by appropriate safety signs. Such signs should comply with the Health and Safety (Safety Signs and Signals) Regulations 1996 or BS 5499-1.[53]

245 To reduce the risk of corrosion, it is sensible to keep extinguishers off the ground and to provide protection against the weather.

246 Extinguishers should be to a recognised standard such as BS EN 3[54] or BS 5423[55] and be suitable for

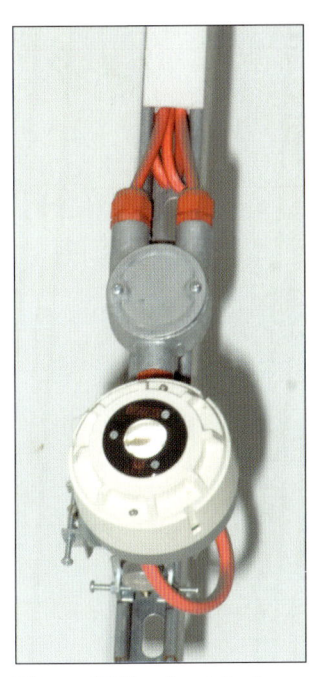

Figure 37 Smoke detector

Figure 38 A fire marshall monitoring the fire alarm system

tackling fires involving the dangerous substances stored. (BS 5423 has now been withdrawn and all new extinguishers should comply with BS EN 3 but existing extinguishers complying with BS 5423 are still acceptable if already in situ and remaining serviceable.) You should seek the advice and guidance of your local fire and rescue authority or equipment supplier on the type and size of fire extinguishers required.

247 There should be an effective means of both raising the alarm and giving warning in case of fire in the storage area (see BS 7974[51] and PD 7974[52]). It should alert all those likely to be affected by the fire. This may vary from small storage areas, where a shout of 'fire' might suffice, to larger areas where a klaxon or siren might be required. You need to discuss your requirements with the fire and rescue authority who will advise on appropriate systems.

248 An assembly point should be identified for people evacuating from such areas, so that they can be accounted for. It should be safe from the effects of fire and smoke. Careful consideration is needed if the smoke can be particularly toxic, eg with fires in pesticide stores, or if there is a risk of flying missiles, such as with aerosol stores. In these cases, the assembly point may be on an alternative site or within another building.

Fire protection

249 Measures such as compartmentation (ie the storage of the packaged dangerous substances in a fire-resistant enclosure) can limit the spread of fire and restrict damage to a specific area. The duration of the protection will depend on the notional period of fire resistance of the enclosure, so if you decide to use this method, the required period of fire resistance will need to be determined. This will depend on a variety of factors including the anticipated fire load and duration, and the time for the Fire and Rescue Service to arrive and start tackling the fire.

Figure 40 Fire extinguishers

Figure 39 Firefighting hoses

Fire suppression systems

250 By tackling a fire almost as soon as it is detected, automatic fire suppression systems can significantly reduce both the risk and damage the fire would otherwise pose if left to develop unchallenged.

251 Where fire suppression systems are installed, it is important, especially in those warehouses where the materials stored frequently change, to ensure the system is appropriate for the contents. The most commonly encountered system is the automatic sprinkler installation, typically using water as the extinguishing medium. However, you should be aware that water is not a suitable extinguishing medium for all fires – it can make some worse. If you are going to install a sprinkler system in a warehouse you need to give serious thought as to what you are likely to store in it.

252 There are two basic types of sprinkler installation:

- **Sealed sprinkler-head system**: each sprinkler head is sealed with a temperature-sensitive device, eg a fusible link or glass bulb. As such, the only devices that operate are those which become sufficiently heated, ie those in the vicinity of the fire. This then limits the discharge of water to that area. These systems are ideal for controlling fires in materials that can be 'wetted', ie absorb water. Also, by limiting water discharge to the area of the fire, water damage to stock is reduced and the size of the containment system needed to prevent firewater run-off is minimised.
- **Open deluge system**: the system is linked to an appropriate fire detection system where, when a fire is detected, water is immediately discharged over an area or zone. This is to wet and cool the materials involved in the fire and anything close-by that might otherwise become involved. Such systems are useful against fires involving highly flammable materials that might be difficult to extinguish once ignited.

253 It is vital to ensure that water from sprinklers does not cause the stored dangerous substances to react as this may escalate the situation. This should form part of your fire risk assessment and further advice can be obtained from the fire and rescue authority. However, when used in appropriate situations sprinklers can be effective by fighting fires using less firefighting water and targeting it where it is required. This results in a smaller volume of contaminated water that must be decontaminated following a fire and can reduce the environmental impact. In previous cases of warehouse fires, the main source of environmental damage was contaminated firefighting water entering the environment in large quantities.

254 You should note that fires involving flammable liquids, especially those immiscible in water, are unlikely to be controlled by water alone. Indeed, it may cause the fire to spread. In some circumstances the use of firefighting foam with a sprinkler system will provide effective protection for stocks of flammable liquid. Foam may not, however, be effective on 'running fires', eg fires in high-racked stores of flammable liquids in plastic containers.

Figure 41 Manifold for a sprinkler system

Figure 42 A firewater tank

255 It is not generally recommended that firewater be recirculated when fighting a fire. Contaminated the water, eg with flammable substances, may make the fire worse.

256 Automatic fire suppression systems should be designed and installed to a required standard, such as BS 5306-2.[56] However, care should be taken in using the commodity classification scheme within the standard. Palletised stocks of flammable liquids in plastic containers should be classified as an 'oil and flammable liquid hazard' rather than 'flammable liquids in combustible containers'. As with all systems required to work on demand, it is imperative that they are correctly maintained and serviced. This is especially true with, for example, a suppression system.

Smoke control systems

257 The discharge of smoke from a building in the early stages of a fire can help protect the means of escape, and also assist the Fire and Rescue Service in their firefighting operations and delay lateral fire spread.

258 The ideal smoke control systems are purpose-designed for the building and the materials stored in it. There are essentially two types: natural ventilation using opening vents or a powered exhaust system that operates at a specified temperature.

259 Glass roof lights, which will shatter in a fire, can also provide effective smoke control, providing they are of sufficient area, as can those made from materials with a low melting point. The latter can, however, drip molten or burning plastic into the building, escalating the fire by acting as a source of ignition or affecting the operation of sprinkler heads. It is also unlikely that they can be made to operate as early in a fire as a purpose-designed system, which is the preferred option.

260 Where a separate fire detection system, or a fire suppression system, is installed, it is important to ensure that it operates before the smoke control system does. The interaction of smoke control systems and fire suppression systems is a complex matter, and the combined system needs to be designed and installed by competent fire protection engineers.

261 More information is available in BS 7346, parts 4[57] and 5.[58]

Emergency procedures

262 Initiating emergency procedures at the earliest stage of an incident can significantly reduce the impact on people, premises and the environment. You need to develop a procedure for dealing with emergencies. Consideration needs to be given to the range of possible events, taking into account the following:

- the nature and quantities of the dangerous substances stored;
- the location of the storage facility and its design;
- the people, both on site and off site, who may be affected; and
- possible environmental impacts.

263 You may have a storage area where any incident is likely to be confined to that area, or to the building containing the store. In this case the emergency procedures may be limited to ensuring that everyone can safely escape from the effects of a fire or toxic gas release, and that the Fire and Rescue Service is called with minimum delay.

264 The Fire and Rescue Service has duties under the Fire and Rescue Services Act 2004 to enable it to tackle any outbreak of fire. This includes familiarising itself with the means of access to premises and the layout, including the availability of water supplies. To assist in this, you should agree the following with your local fire and rescue authority:

- the provision and maintenance of suitable access for firefighting personnel and their vehicles; and
- as necessary, the provision of a convenient fire main and hydrant.

265 Where there is a possibility that a fire in the store might spread to affect other parts, whether on site or off site, you need to consider how the risk to anyone present can be reduced. Similarly, if a fire could reach the store, preventive measures have to be considered.

266 Where you conclude, in consultation with your local fire and rescue authority, that precautions are needed, the extent will depend on the nature of the site. They could vary from housing suitable fire extinguishers or fire hose reels to tackle an incipient fire, to installing sprinkler systems.

267 People expected to use the equipment need to be trained and rehearsed in how to do so, without exposing themselves or others to any unnecessary risk from the fire. This needs to be discussed with the fire and rescue authority.

268 Upon arrival, the Fire and Rescue Service will assume responsibility for firefighting operations. It is therefore important that they are aware of the firefighting equipment and capability on site. This includes having in place agreed procedures with the works fire team (if there is one) to ensure that control of the incident is maintained and that nobody is exposed to unnecessary risk during handover. Any subsequent role for the works fire team should be agreed with the Fire and Rescue Service and detailed in your emergency procedures.

269 It is recommended that, where a number of different types of dangerous substances are stored, an inventory of the stock be readily available. This record should provide details of the quantity and location of all the dangerous substances in the store. The record will require updating to take account of stock movements or at the end of the working day. A copy of the record should be available at a point on the site which is unlikely to be affected by an emergency, so it can be used by both management and the emergency services when dealing with an incident.

270 Where 25 tonnes or more of dangerous substances are stored, DSEAR will apply. These Regulations make specific requirements for posting hazard warning signs and for the design of the signs to be used. You should consult the fire and rescue authority about their requirements for the actual siting of the signs.

271 It would be useful if the emergency services were given an out-of-hours telephone contact number so that they can obtain specialist advice when dealing with an incident.

Control of off-site risks

272 Firewater run-off is often highly polluting and may also place a major strain on normal drainage facilities. Allowance for firewater can be made in bunded storage areas. However, additional containment systems, such as firewater lagoons, interceptor pits or tanks, may be necessary. This is particularly so at large installations and those storing very polluting substances, where there is a risk of contaminating local watercourses, groundwater and soil. Where there is a risk of pollution from firewater run-off (this should be considered during the risk assessment), you should consult with the Environment Agency (or SEPA in Scotland) and your local fire and rescue authority.

273 Where procedures state that firefighting water is to be contained on site, a good management system should be in place to ensure rainwater etc does not build up within the containment systems rendering them inadequate in the event of a fire. When planning such systems, note that firefighters tend to use water in abundance. Use can typically range from 1000 m^3 to 20 000 m^3. Underground storage systems can be effective as they limit oxygen exposure to possibly contaminated firewater and can include carbon dioxide flooding systems to extinguish surface fires. This type of arrangement is generally restricted to larger sites. You should consult with the fire and rescue authority for guidance on what measures are suitable for your chemical warehouse.

274 Guidance on this topic can be found in *Design of containment systems for prevention of water pollution from industrial incidents*.[59] Further guidance is also available from the Environment Agency in *Managing fire water and major spillages* PPG18.[60]

275 Where foreseeable incidents may affect people, property or the environment beyond the site boundary, the emergency services should be consulted when preparing the emergency procedures. Such discussions should include firefighting strategies, including the adoption of a controlled burn to protect people and the environment.

276 If your site is subject to regulations 7–14 of the COMAH Regulations, you are required to produce on-site and off-site emergency plans. Further guidance is available in *Emergency planning for major accidents. Control of Major Accident Hazards Regulations 1999 (COMAH)* HSG191.[61]

Escape facilities

277 DSEAR requires that, where the risk assessment indicates, escape facilities be provided and maintained to ensure that in the event of danger people can leave places quickly and safely. Means of escape in case of fire constitute part of the general fire precautions and are subject to the relevant legislation referred to in paragraph 239. The hazardous properties of the stored substances should be taken into account when planning escape facilities.

First aid

278 The Health and Safety (First-Aid) Regulations 1981 require you to provide adequate and appropriate equipment, facilities and personnel to enable first aid to be given to your employees if they become ill or injured while at work.

279 You will need to consider what dangerous substances and activities are on site to determine your first-aid provisions. The material safety data sheets for your dangerous substances will help you to determine what provisions are required. This will form part of your risk assessment.

280 Further guidance is available in *First-aid at work: Your questions answered* INDG214.[62]

Figure 43 Instructions to initiate the emergency procedure

Information, instruction and training

281 Failures in training, operating procedures and supervision have been shown to be among the root cause of many incidents, some very serious. So if employees are to make a maximum contribution to health and safety, there should be proper arrangements in place to ensure they are competent. This is more than simply training them, as experience of applying skills and knowledge gained under supervision is also required.

282 Health and safety legislation requires that training be provided to ensure people are competent to undertake their duties at work. Specifically, DSEAR requires that employees be given training to safeguard themselves from the dangerous substances on site. Training should also be provided in the use and application of control and mitigation measures, and equipment that is used on site, taking into account the recommendations and instructions supplied by the manufacturer.

283 Depending on the nature of your activity you may be required by regulation 43(3) of the Carriage Regulations to appoint a safety advisor to comply with the requirements relating to duties of safety advisors defined in ADR.[15] You may also wish to consider the training needs of such an advisor.

284 You need to take into account the needs of people other than employees, eg contractors and visitors, who may be present on site. It may not always be possible to provide formal training; however, you need to consider what risks these people will be exposed to on your site and take steps to reduce them. Information, instruction and training on the dangerous substances on site may only be required for non-employees where it is necessary to ensure their safety. However, you will need to consider factors such as duration of stay on your site and what duties (if any) they will be undertaking while in your chemical warehouse. It may be that some form of formal training is deemed necessary for contractors.

285 Information, instruction and training will need to be reviewed periodically and, if necessary, revised if there is any significant change to the type of work done or dangerous substances stored on site. This includes changes to the site's inventory of dangerous substances, control and mitigation measures or automation of processes etc.

286 Proper consultation with those who do the work is crucial in helping to raise awareness of health and safety and environmental protection. Employers should consult their employees and their representatives in accordance with the Health and Safety (Consultation with Employees) Regulations 1996 and the Safety Representatives and Safety Committees Regulations 1977.

287 HSE statistics show a direct link between the presence of a workplace safety representative and increased awareness of health and safety issues on site. Competent representatives can make effective contributions by participating in hazard spotting, problem solving and investigation initiatives. This can result in a lower injury rate, better working practices, reduced costs, and greater workforce participation and consultation.

288 If your chemical warehouse is subject to the COMAH Regulations, you are required to consult your employees or their representatives about the preparation of an on-site emergency plan. You may still need to consider, particularly for smaller sites, whether there is a need for professional health and safety advice from within or outside the chemical warehouse to maintain or achieve a high level of competence.

289 Further guidance on these issues can be obtained from the DSEAR Approved Codes of Practice,[8–12] *A Guide to the Control of Major Accident Hazards Regulations 1999 (as amended): Guidance on Regulations* L111[5] and *Successful health and safety management* HSG65.[63]

Audit and review

Audit

290 An audit is defined as:

'the structured process of collecting independent information on the efficiency, effectiveness and reliability of the total health and safety management system and drawing up plans for corrective action.'

291 All risk control systems deteriorate over time so auditing will help you assess whether your health and safety management system is still effective. A comprehensive picture of how effectively the health and safety management system within the chemical warehouse is controlling the risks will emerge from a well-structured auditing programme indicating when and how each component part will be audited. The adequacy of your system can then be assessed against a relevant 'standard'. Audits should be conducted periodically. Further guidance is available in *Successful health and safety management* HSG65.[63]

292 Some recent events in the UK and elsewhere have shown that audit processes have vulnerabilities. It is not enough to have processes in place. Audits should be capable of detecting non-compliance and providing managers with useful data to improve performance.

Reviewing performance

293 Reviewing is the process of making judgements about the adequacy of performance and taking decisions about the nature and timing of the actions necessary to remedy deficiencies. The main sources of information come from measuring activities and audits, and reviewing should be a continuous process undertaken at different levels within your organisation. A small number of carefully chosen indicators can monitor the status of key risk control systems and provide an early warning should controls deteriorate dangerously. This is particularly important for sites containing an inventory of dangerous substances with the potential for a major incident, such as chemical warehouses.

294 There are two types of process safety performance indictors used on sites with dangerous substances. They are known as leading and lagging indicators:

- Leading indicators are a form of active monitoring focused on a critical risk control system to ensure its continued effectiveness. **Leading indicators require a routine systematic check that key actions or activities are undertaken as intended.** They can be considered as measures of process or inputs essential to deliver the desired safety outcome. Examples of leading indicators are the fraction of maintenance actions identified that are completed within a specified time or the fraction of safety-critical equipment that performs to specification when inspected or tested. As with audits, indicators should provide data that can be used to improve performance. Managers should be able to show how the indicators are used for this purpose.
- Lagging indicators are a form of reactive monitoring requiring the reporting and investigation of specific incidents and events to discover weaknesses in that system. These incidents or events may not have to result in major damage, injury or loss of containment, providing that they represent a failure of a significant control system which guards against or limits the consequences of a major incident. **Lagging indicators show when a desired safety outcome has failed or has not been achieved.** Examples of lagging indicators are rates of accidents or dangerous occurrences, or the number of unexpected loss-of-containment incidents.

295 Monitoring the performance of management systems intended to control or mitigate major hazard risks using leading and lagging indicators is considered good practice at COMAH sites. Guidance on setting performance indicators is available in *Developing process safety indicators: A step-by-step guide for chemical and major hazard industries* HSG254.[64] Further sector-specific guidance for warehouse operations will be developed by the relevant trade associations supported by HSE by the end of 2009.

References and further reading

References

1 *Managing contractors: A guide for employers. An open learning booklet* HSG159 HSE Books 1997 ISBN 978 0 7176 1196 6

2 *Storage & handling of drums & intermediate bulk containers* Pollution Prevention Guidelines PPG26 Environment Agency 2004

3 *Getting Your Site Right: Industrial and Commercial Pollution Prevention* Environment Agency 2004

4 *Warehousing and storage: A guide to health and safety* HSG76 (Second edition) HSE Books 2007 ISBN 978 07176 6225 8

5 *A guide to the Control of Major Accident Hazards Regulations 1999 (as amended). Guidance on Regulations* L111 HSE Books 2006 ISBN 978 0 7176 6175 6

6 *The storage of flammable liquids in containers* HSG51 (Second edition) HSE Books 1998 ISBN 978 0 7176 1471 4

7 *Notification and marking of sites. The Dangerous Substances (Notification and Marking of Sites) Regulations 1990. Guidance on Regulations* HSR29 HSE Books 1990 ISBN 978 0 11 885435 1

8 *Design of plant, equipment and workplaces. Dangerous Substances and Explosive Atmospheres Regulations 2002. Approved Code of Practice and guidance* L134 HSE Books 2003 ISBN 978 0 7176 2199 6

9 *Storage of dangerous substances. Dangerous Substances and Explosive Atmospheres Regulations 2002. Approved Code of Practice and guidance* L135 HSE Books 2003 ISBN 978 0 7176 2200 9

10 *Control and mitigation measures. Dangerous Substances and Explosive Atmospheres Regulations 2002. Approved Code of Practice and guidance* L136 HSE Books 2003 ISBN 978 0 7176 2201 6

11 *Safe maintenance, repair and cleaning procedures. Dangerous Substances and Explosive Atmospheres Regulations 2002. Approved Code of Practice and guidance* L137 HSE Books 2003 ISBN 978 0 7176 2202 3

12 *Dangerous substances and explosive atmospheres. Dangerous Substances and Explosive Atmospheres Regulations 2002. Approved Code of Practice and guidance* L138 HSE Books 2003 ISBN 978 0 7176 2203 0

13 *Memorandum of guidance on the Electricity at Work Regulations 1989. Guidance on Regulations* HSR25 (Second edition) HSE Books 2007 ISBN 978 0 7176 6228 9

14 *Management of health and safety at work. Management of Health and Safety at Work Regulations 1999. Approved Code of Practice and guidance* L21 (Second edition) HSE Books 2000 ISBN 978 0 7176 2488 1

15 *ADR 2009: European Agreement concerning the International Carriage of Dangerous Goods by Road* (ADR) UN, ECE 2008 ISBN 978 92 1 139131 2 www.unece.org/trans/danger/publi/adr/adr2009/09ContentsE.html (ADR is updated every two years. This 2009 edition will be mandatory from 1 July 2009)

16 BS EN 1089-3:2004 *Transportable gas cylinders. Gas cylinder identification (excluding LPG). Colour coding* British Standards Institution

17 *Storing and handling ammonium nitrate* Leaflet INDG230 HSE Books 1996 www.hse.gov.uk/pubns/indg230.pdf

18 *Storage and use of sodium chlorate and other similar strong oxidants* Chemical Safety Guidance Note CS3 (Fourth edition) HSE Books 1998 ISBN 978 0 7176 1500 1

19 *The storage and handling of organic peroxides* Chemical Safety Guidance Note CS21 HSE Books 1991 ISBN 978 0 7176 2403 4

20 *Five steps to risk assessment* Leaflet INDG163(rev2) HSE Books 2006 (single copy free or priced packs of 10 ISBN 978 0 7176 6189 3) www.hse.gov.uk/pubns/indg163.pdf

21 *Storage of Full and Empty LPG Cylinders and Cartridges* Code of Practice 7 LP Gas Association 2004 www.lpga.co.uk

22 *Energetic and spontaneously combustible substances: Identification and safe handling* HSG131 HSE Books 1995 ISBN 978 0 7176 0893 5

23 *Safe use and storage of cellular plastics* HSG92 HSE Books 1996 ISBN 978 0 7176 1115 7

24 *Storage and handling of industrial nitrocellulose* HSG135 HSE Books 1995 ISBN 978 0 7176 0694 8

25 *Building Regulations 2000 Approved Document B: Fire Safety* (Incorporating 2000 and 2002 amendments) The Stationery Office 2006 ISBN 978 0 11 703634 5

26 The *Building (Scotland) Regulations 2004* www.sbsa.gov.uk

27 BS 7671:2008 *Requirements for electrical installations. IEE Wiring Regulations. Seventeenth edition* British Standards Institution

28 BS EN 60079-14:2003 *Electrical apparatus for explosive gas atmospheres. Electrical installations in hazardous areas (other than mines)* British Standards Institution

29 *Lift trucks in potentially flammable atmospheres* HSG113 HSE Books 1996 ISBN 978 0 7176 0706 8

30 PD CLC/TR 50404:2003 *Electrostatics: Code of practice for the avoidance of hazards due to static electricity* British Standards Institution

31 *Safe working with industrial solvents: Flammability: A safety guide for users* Best Practice Guidelines No 4 European Solvents Industry Group 2003 www.esig.org

32 *'The safe use of IBCs with flammable liquids' Loss Prevention Bulletin* issue 177 pages 3-8

33 *The use of IBCs for oxygenated and hydrocarbon solvents* Guidance notice No 51a Solvents Industry Association 2003 www.sia-uk.org.uk

34 *Guidance on permit-to-work systems: A guide for the petroleum, chemical and allied industries* HSG250 HSE Books 2005 ISBN 978 0 7176 2943 5

35 *The BAMA Standard for Consumer Safety and Good Manufacturing Practice: Module 6 Warehousing* Issue 3 British Aerosol Manufacturers' Association 2007

36 *Guidance for the storage of liquids in intermediate bulk containers* CBA/SIA 2008 www.chemical.org.uk

37 *Groundwater Protection Code: Solvent use & storage* PB 9849 Defra 2004

38 *Proposed Environment Agency, SEPA, NI EHS and HSE joint guidance on the Storage of Hazardous Wastes*, available at www.environment-agency.gov.uk

39 BS EN 62305:2006 *Protection against lightning* British Standards Institution

40 *Control of substances hazardous to health (Fifth edition). The Control of Substances Hazardous to Health Regulations 2002 (as amended). Approved Code of Practice and guidance* L5 (Fifth edition) HSE Books 2005 ISBN 978 0 7176 2981 7

41 BS 5415-2.2:Supplement No. 1:1986 *Safety of electrical motor-operated industrial and commercial cleaning appliances. Particular requirements. Specification for type H industrial vacuum cleaners for dusts hazardous to health* British Standards Institution

42 *Personal protective equipment at work (Second edition). Personal Protective Equipment at Work Regulations 1992 (as amended). Guidance on Regulations* L25 (Second edition) HSE Books 2005 ISBN 978 0 7176 6139 8

43 BS EN 60079-10:2003 *Electrical apparatus for explosive gas atmospheres. Classification of hazardous areas* British Standards Institution

44 BS EN 50281-3:2002 *Electrical apparatus for use in the presence of combustible dust: Classification of areas where combustible dusts are or may be present* British Standards Institution

45 *Area Classification Code for Installations Handling Flammable Fluids (Model Code of Safe Practice in the Petroleum Industry)* Energy Institute 2005 ISBN 978 0 85293 418 0

46 *Joint CBA and SIA guidance for the storage of flammable liquids in sealed packages in specified external storage areas* CBA/SIA 2005 Available from CBA website www.chemical.org.uk

47 *Safety Report Assessment Guide: Chemical warehouses – Hazards* Version 6 HSE 2002 www.hse.gov.uk/comah/sragcwh

48 *Guide to Fire Safety in Factories and Warehouses (Fire Safety Employers' Guide)* Communities and Local Government 2006 ISBN 978 1 85112 816 7

49 *Practical fire safety guidance for factories and storage premises* Scottish Executive

50 BS 5839-1:2002 *Fire detection and fire alarm systems for buildings. Code of practice for system design, installation, commissioning and maintenance* British Standards Institution

51 BS 7974:2001 *Application of fire safety engineering principles to the design of buildings. Code of practice* British Standards Institution

52 PD 7974-1:2003 *Applications of fire safety engineering principles to the design of buildings. Initiation and development of fire within the enclosure of origin (Sub-system 1)* British Standards Institution

53 BS 5499-1:2002 *Graphical symbols and signs. Safety signs, including fire safety signs. Specification for geometric shapes, colours and layout* British Standards Institution

54 BS EN 3 *Portable fire extinguishers* (7 Parts) British Standards Institution

55 BS 5423:1987 *Specification for portable fire extinguishers* British Standards Institution (Superseded, withdrawn)

56 BS 5306-2:1990 *Fire extinguishing installations and equipment on premises. Specification for sprinkler systems* British Standards Institution

57 BS 7346-4:2003 *Components for smoke and heat control systems. Functional recommendations and calculation methods for smoke and heat exhaust ventilation systems, employing steady-state design fires* British Standards Institution

58 BS 7346-5:2005 *Components for smoke and heat control systems. Functional recommendations and calculation methods for smoke and heat exhaust ventilation systems, employing time-dependent design fires* British Standards Institution

59 *Design of containment systems for prevention of water pollution from industrial incidents* R164 Construction Industry Research and Information Association 1996 ISBN 978 0 86017 476 9

60 *Managing fire water and major spillages* Pollution Prevention Guidelines PPG18 Environmental Agency

61 *Emergency planning for major accidents: Control of Major Accident Hazards Regulations 1999 (COMAH)* HSG191 HSE Books 1999 ISBN 978 0 7176 1695 4

62 *First aid at work: Your questions answered* Leaflet INDG214 HSE Books 1997 (single copy free or priced packs of 15 ISBN 978 0 7176 1074 7) www.hse.gov.uk/pubns/indg214.pdf

63 *Successful health and safety management* HSG65 (Second edition) HSE Books 1997 ISBN 978 0 7176 1276 5

64 *Developing process safety indicators: A step-by-step guide for chemical and major hazard industries* HSG254 HSE Books 2006 ISBN 978 0 7176 6180 0

Further reading

Code of Practice for suppliers of pesticides to agriculture, horticulture and forestry Ministry of Agriculture, Fisheries and Food 1997 (The Yellow Code)

EH40/2005 Workplace exposure limits: Containing the list of workplace exposure limits for use with the Control of Substances Hazardous to Health Regulations 2002 (as amended) Environmental Hygiene Guidance Note EH40 HSE Books 2005 ISBN 978 0 7176 2977 0

Fire performance of composite IBCs RR564 HSE Books 2007 www.hse.gov.uk/research/rrhtm/index.htm

Guidance for the recovery and disposal of hazardous and non-hazardous waste Sector Guidance Note IPPC S5.06 Environment Agency 2001

Guidelines for Safe Warehousing: Guidelines for Safe Warehousing of Substances with Hazardous Characteristics Chemical Industries Association 1990

Guidance for the storage of transportable gas cylinders for industrial use GN2 (Revision 3) British Compressed Gases Association 2005

Guidance on storing pesticides for farmers and other professional users Agriculture Information Sheet AIS16 HSE Books 1996 www.hse.gov.uk/pubns/agindex.htm

IP Model Code of Safe Practice: Area Classification Code for Installations – Handling Flammable Fluids Pt. 15 (Model Code of Safe Practice in the Petroleum Industry) Energy Institute 2002 ISBN 978 0 85293 223 0

Reducing error and influencing behaviour HSG48 (Second edition) HSE Books 1999 ISBN 978 0 7176 2452 2

Responsible Care Management Systems Guidance (Fourth edition) RC127 Chemical Industries Association 2003

Safety signs and signals. The Health and Safety (Safety Signs and Signals) Regulations 1996. Guidance on Regulations L64 HSE Books 1996 ISBN 978 0 7176 0870 6

Relevant legislation

Building Regulations 2000 SI 2000/2531 The Stationery Office 2000 ISBN 978 0 11 099897 9

Building (Scotland) Regulations 2004 SI 2004/406 The Stationery Office 2004 ISBN 978 0 11 069246 7

Carriage of Dangerous Goods and Use of Transportable Pressure Equipment Regulations 2007 SI 2007/1573 The Stationery Office 2007 ISBN 978 0 11 077469 5

Chemicals (Hazard Information and Packaging for Supply) Regulations 2002 SI 2002/1689 The Stationery Office 2002 ISBN 978 0 11 042419 4 (as amended)

CLP regulations: 'Regulation (EC) No 1272/2008 of the European Parliament and of the Council of 16 December 2008 on classification, labelling and packaging of substances and mixtures, amending and repealing Directives 67/548/EEC and 1999/45/EC, and amending Regulation (EC) No 1907/2006' *Official Journal of the European Union* 31.12.2008 **51** L353/1

Control of Major Accident Hazards Regulations 1999 SI 1999/743 The Stationery Office 1999 ISBN 978 0 11 082192 4 (as amended)

Control of Pesticides Regulations 1986 SI 1986/1510 The Stationery Office 1986 ISBN 978 0 11 067510 7 (as amended)

Control of Pollution Act 1974 (c. 40) The Stationery Office 1974 ISBN 978 0 10 544074 1

Control of Substances Hazardous to Health Regulations 2002 SI 2002/2677 The Stationery Office 2002 ISBN 978 0 11 042919 9 (as amended)

Dangerous Substances and Explosive Atmospheres Regulations 2002 SI 2002/2776 ISBN 978 0 11 042957 1

Dangerous Substances (Notification and Marking of Sites) Regulations 1990 SI 1990/304 The Stationery Office 1990 ISBN 978 0 11 003304 4

Electricity at Work Regulations 1989 SI 1989/635 The Stationery Office 1989 ISBN 978 0 11 096635 9

Environmental Protection Act 1990 (c. 43) The Stationery Office 1990 ISBN 978 0 10 544390 2

Equipment and Protective Systems Intended for Use in Potentially Explosive Atmospheres Regulations 1996 SI 1996/192 The Stationery Office 1996 ISBN 978 0 11 053999 7 (as amended)

Fire and Rescue Services Act 2004 (c. 21) The Stationery Office 2004 ISBN 978 0 10 542104 7

Fire (Scotland Act) 2005 (c. 5) The Stationery Office 2005 ISBN 978 0 10 590078 8

Fire Services Act 1947 (c. 41) The Stationery Office 1947 ISBN 978 0 10 850109 8

Food and Environmental Protection Act 1985 (c. 48) The Stationery Office 1985 ISBN 978 0 10 544885 3 Part 3

Health and Safety at Work etc Act 1974 (c. 37) The Stationery Office 1974 ISBN 978 0 10 543774 1

Health and Safety (Consultation with Employees) Regulations 1996 SI 1996/1513 The Stationery Office 1996 ISBN 978 0 11 054839 5

Health and Safety (Enforcing Authority) Regulations 1998 SI 1998/494 The Stationery Office 1998 ISBN 978 0 11 065642 7

Health and Safety (First-Aid) Regulations 1981 SI 1981/917 The Stationery Office 1981 ISBN 978 0 11 016917 0

Health and Safety (Safety Signs and Signals) Regulations 1996 SI 1996/341 The Stationery Office 1996 ISBN 978 0 11 054093 1

Ionising Radiations Regulations 1999 SI 1999/3232 The Stationery Office 1999 ISBN 978 0 11 085614 8

Management of Health and Safety at Work Regulations 1999 SI 1999/3242 The Stationery Office 1999 ISBN 978 0 11 085625 4 (as amended)

Notification of Installations Handling Hazardous Substances Regulations 1982 SI 1982/1357 The Stationery Office 1982 ISBN 978 0 11 027357 0

Planning (Control of Major-Accident Hazards) Regulations 1999 SI 1999/981 The Stationery Office 1999 ISBN 978 0 11 082367 6

Planning (Hazardous Substances) Act 1990 (c. 10) The Stationery Office 1990 ISBN 978 0 10 541090 4

Planning (Hazardous Substances) Regulations 1992 SI 1992/656 The Stationery Office 1992 ISBN 978 0 11 023656 8

Provision and Use of Work Equipment Regulations 1998 SI 1998/2306 The Stationery Office 1998 ISBN 978 0 11 079599 7

Regulatory Reform (Fire Safety) Order 2005 SI 2005/1541 The Stationery Office 2005 ISBN 978 0 11 072945 9

Safety Representatives and Safety Committees Regulations 1977 SI 1977/500 The Stationery Office 1977 ISBN 978 0 11 070500 2

Water Resources Act 1991 (c. 57) The Stationery Office 1991 ISBN 978 0 10 545791 6

Workplace (Health, Safety and Welfare) Regulations 1992 SI 1992/3004 The Stationery Office 1992 ISBN 978 0 11 025804 1 (as amended)

Glossary

ACOP Approved Code of Practice.
ADR Accord dangereux routier (European agreement concerning the international carriage of dangerous goods by road).
ALARP as low as reasonably practicable.
ATEX Atmosphere Explosiv (used in the context of two European Directives, 94/9/EC and 1999/92/EC).
AZDN azodiisobutyronitrile – more correctly, azobisisobutyronitrile. A toxic and unstable organic solid that can explode if subjected to heat or shock.

BAMA British Aerosol Manufacturers' Association.

Carriage Regulations Carriage of Dangerous Goods and Use of Transportable Pressure Equipment Regulations 2007 (due to be replaced in 2009 by Carriage Regulations 2009).
CBA Chemical Business Association.
CHIP Chemicals (Hazard Information and Packaging for Supply) Regulations 2002.
CIMAH Control of Industrial Major Accident Hazards Regulations 1984.
COMAH Control of Major Accidents Hazards Regulations 1999.
containers all receptacles, packages and packaging.
COSHH Control of Substances Hazardous to Health Regulations 2002.

DSEAR Dangerous Substances and Explosive Atmospheres Regulations 2002.

EPS Regulations Equipment and Protective Systems Intended for Use in Potentially Explosive Atmospheres Regulations 1996.

GHS Globally Harmonised System of classification and labelling of chemicals. A UN System to identify hazardous chemicals and inform users about these hazards through standard symbols and phrases on packaging labels and through MSDSs.

IBC intermediate bulk container.
IPPC integrated pollution prevention and control.

LPG liquefied petroleum gas.

Management Regulations the Management of Health and Safety at Work Regulations 1999, as amended by the Management of Health and Safety at Work (Amendment) Regulations 2006.
MSDS material safety data sheet.

NIEHS Northern Ireland Environment and Heritage Service, now Northern Ireland Environment Agency.
NIHHS Notification of Installations Handling Hazardous Substances Regulations.

OFCOM Office of Communications, the independent regulator and competition authority for the UK communications industries.

PCB polychlorinated biphenyl.
PPE personal protective equipment.

REACH Registration, Evaluation, Authorisation and Restriction of Chemicals.
RF radio frequency.

SEPA Scottish Environmental Protection Agency.
SIA Solvents Industry Association.

Further information

HSE priced and free publications are available by mail order from HSE Books, PO Box 1999, Sudbury, Suffolk CO10 2WA Tel: 01787 881165 Fax: 01787 313995 Website: www.hsebooks.co.uk (HSE priced publications are also available from bookshops and free leaflets can be downloaded from HSE's website: www.hse.gov.uk.)

For information about health and safety ring HSE's Infoline Tel: 0845 345 0055 Fax: 0845 408 9566 Textphone: 0845 408 9577 e-mail: hse.infoline@natbrit.com or write to HSE Information Services, Caerphilly Business Park, Caerphilly CF83 3GG.

British Standards can be obtained in PDF or hard copy formats from the BSI online shop: www.bsigroup.com/Shop or by contacting BSI Customer Services for hard copies only Tel: 020 8996 9001 e-mail: cservices@bsigroup.com.

The Stationery Office publications are available from The Stationery Office, PO Box 29, Norwich NR3 1GN Tel: 0870 600 5522 Fax: 0870 600 5533 e-mail: customer.services@tso.co.uk Website: www.tso.co.uk (They are also available from bookshops.) Statutory Instruments can be viewed free of charge at www.opsi.gov.uk.

For further information on publications from the Environment Agency or the Scottish Environment Protection Agency (SEPA), refer to their websites www.environment-agency.gov.uk and www.sepa.org.uk.

Chemical warehousing